kabnf KO
609.7 KRIM
Krim, Robert M., author
Boston made
33410017044118   03-17-2021

Kouts Public Library
101 E. Daumer Road
Kouts, IN 46347

BOSTON MADE

BOSTON MADE

FROM REVOLUTION TO ROBOTICS—
INNOVATIONS THAT CHANGED THE WORLD

Dr. Robert M. Krim
with Alan R. Earls

Copyright © 2021 by Robert M. Krim and Alan R. Earls

All rights reserved, including the right of reproduction in whole or in part in any form. Charlesbridge and colophon are registered trademarks of Charlesbridge Publishing, Inc.

At the time of publication, all URLs printed in this book were accurate and active. Charlesbridge and the author are not responsible for the content or accessibility of any website.

An Imagine Book
Published by Charlesbridge
9 Galen Street
Watertown, MA 02472
(617) 926-0329
www.imaginebooks.net

Library of Congress Cataloging-in-Publication Data
Names: Krim, Robert M., author. | Earls, Alan R., author.
Title: Boston made : from revolution to robotics—innovations that changed the world / by Dr. Robert M. Krim, with Alan R. Earls.
Description: Watertown : Charlesbridge Publishing, [2021] | Summary: "An illustrated exploration of how the Greater Boston area became one of the world's leading centers for innovation"— Provided by publisher.
Identifiers: LCCN 2020021602 (print) | LCCN 2020021603 (ebook) | ISBN 9781623545352 (hardcover) | ISBN 9781632892256 (ebook)
Subjects: LCSH: Inventions—Boston Metropolitan Area—History. | Industries—Boston Metropolitan Area—History. | Inventors—United States—Biography. | Boston (Mass.)—Biography.
Classification: LCC T75.B66 K75 2021 (print) | LCC T75.B66 (ebook) | DDC 609.744/61—dc23
LC record available at https://lccn.loc.gov/2020021602
LC ebook record available at https://lccn.loc.gov/2020021603

Printed in China
(hc) 10 9 8 7 6 5 4 3 2 1

Cover design by Cathleen Schaad
Interior design by Jeff Miller

Contents

Introduction 1
    BUMP AND CONNECT 23
    CREATING A BETTER WORLD 27
    ENHANCING PUBLIC LIFE 38
    MATTERS OF LIFE AND DEATH 52
    ENHANCING QUALITY OF LIFE 66
    DOING BUSINESS TAKES MONEY 84
    NINETEENTH-CENTURY HIGH TECH 91
    THAT'S ENTERTAINMENT 102
    CONNECTING PEOPLE 110
    BRAVE NEW WORLD 122
    FINISH WITH A SMILE 135

Conclusion 137
Bibliography 139
Illustration and Photo Credits 142
Index 145

About the Authors 151

Acknowledgments

This book is for my inventor father, Norman Krim, who piqued my innovation interest; my mother, who loved Boston history; my supportive wife, Kathlyne Anderson; and our enterprising next generation, Sarah and Benjamin, who all supported me with the space to focus on Boston's innovation story.

For grasping the importance of the Greater Boston innovation narrative and making it engrossing while they guided this book to publication at Charlesbridge: editor Kevin Stevens and publisher Mary Ann Sabia.

Alan Earls, my erudite coauthor and master of technological history.

Janey Bishoff for decades of incisive contributions to Boston's innovation story and our exhibits at Logan Airport. Inspiring and working on the key questions of Boston and innovation: Steve Crosby, Jim Rooney, Janice Bourque, Steve Grossman, Henry Lee, Phillip Clay, Jill Lepore, Roger Berkowitz, Bill Nigreen, David Feigenbaum, Andrea Cabral, Paul Gray, Richard Freeland, and Anne Bailey Berman. Advisors and scholars: Bob Allison, Ed Glaeser, Westy Egmont, Governor Michael Dukakis, Tim Rowe, the late Jon Lipsky, journalist Scott Kirsner, Massachusetts Historical

Society's Gavin Kleespies and Peter Drummey, SynecticsWorld's Joseph Gammal, and Arthur Krim.

For their hard work on the Innovation Trail & Odyssey, which laid the foundation for understanding innovation: David Bartone, Jenna Leventhal, Meaghan Smith Young, Colin Rowan, David Sears, Dr. Kerri Greenridge, as well as hundreds of great interns over a dozen years.

For a few of the many who played turning-point roles: Tom Paine, for his parallel research on hundreds of Boston innovations; Massport's former CEO Tom Glynn, for commissioning the permanent exhibit "From Massachusetts to the World: Four Centuries of Innovation"; and the Massachusetts Technology Collaborative, for funding a key study.

—Dr. Bob Krim

Introduction

Imagine melted candy leading to one of the greatest modern innovations. So it was for Percy Spencer, who walked through a radar test room at the Boston-area Raytheon Company one day in 1946 and discovered that the chocolate bar in his back pocket had been turned into a sticky mess. A self-taught engineer and quick thinker, Spencer wondered if the same microwave-emitting tubes known as magnetrons, which had been used for military radar (and put Raytheon on the map during World War II), could cook up other snacks. So he brought in a bag of popping corn and put it in front of the magnetron tubes. Presto, he had popcorn. Within a year, Raytheon was selling the Radarange, later known generically as the microwave oven, and over the next forty years the company, and its Amana subsidiary, produced millions of microwave ovens.

It's a great story. But it's only one of nearly five hundred innovations at the heart of Boston's history. No other American city has identified so many world-changing innovations as part of its narrative. Another large American city recently listed all of its firsts—people and innovations—and came up with eighty. Over the past four centuries, Boston and Massachusetts have continually been a center for the inspiration, creation, and development of some five hundred life-changing discoveries.

This midsized city, perched on the northeastern edge of the continent beside the Atlantic Ocean and founded by Puritan settlers, is still dotted with cobblestone streets and costumed re-enactors who passionately tell the story of how the American Revolution began in Boston. It is a place devoted to its sports teams, its prestigious universities, and its esteemed medical centers. Across the centuries, the capital of Massachusetts has nurtured entrepreneurial spirit; encouraged innovation; and rocked the world with medical breakthroughs, technological advancements, and social innovations.

The Boston region witnessed the world's first successful organ transplant. It was the first home to the telephone, modern venture capital, and the creation of the best available medication for multiple sclerosis. The chocolate chip cookie was born here, as were basketball and the first public park in the United States. The state of Massachusetts was the first to abolish slavery and the first to declare marriage equality for gays. Such a broad gamut of innovative events is remarkable and unprecedented.

Of those many hundreds of innovations that have emerged from the Greater Boston region over the last four hundred years and changed the nation or the world, *Boston Made* chooses to tell the stories of fifty of them. Each story is complete in itself—but taken together, they provide an understanding of why the Boston area has been notably innovative over time. This history is not an accident: as you will see, detailing the crucial innovations and examining the common features driving them illustrate clearly how Boston is a case study in innovative excellence that provides potential lessons for other regions seeking to become or remain vibrant and relevant in a rapidly changing world. (Boston, long the dominant city in Massachusetts and New England, is used throughout this book to describe the specific geographic entity itself as well as the region that surrounds it.)

Boston Made details how successive waves of innovation have allowed Boston to reinvent itself repeatedly, remake its economy, and retain its relevance. A deep dive into these innovation stories reveals the reasons for this excellence. Innovation always has a little magic about it, a little something that evades definition. But that magic doesn't happen in a vacuum. Every one of the innovations described in this book, to varying degrees, emerged from a perfect storm of circumstance, a context that allowed startling and unlikely successes to flow. And that context is made up of five key drivers: strong entrepreneurship, local networking, local funding, local demand, and global demand. Examining these interrelated drivers, which we will do later in this introduction, reveals the general characteristics that transcend the unique circumstances of Boston's history and illuminates the challenges and opportunities faced by other cities and regions. In other words, the Boston story holds lessons that can be applied anywhere that people seek to change their city and world for the better.

But first, what do we mean by the word *innovation*? Certainly, it is important to have great thinkers and researchers, especially Nobel Prize winners like Dr. Joseph Murray, who performed the world's first successful organ transplant, or the diplomat Ralph Bunche, in 1950 the first African American recipient of the Nobel Peace Prize for his mediation of the Arab–Israeli conflict. But the important thing for Boston and the world is that there are people and groups who think great thoughts and then change them into world-altering actions. The invention itself is important, but this book emphasizes why and how each innovation came about and then gained traction so it could become meaningfully used or universally known. Patents and inventions are important, but innovations change lives. So the date we attach to each innovation is not necessarily when it was first conceived but rather when it became commercially or culturally important or when real change occurred because of it. For the most part, *Boston Made*, detailing fifty out of hundreds of innovations, can be seen as a companion to "Four Centuries of Innovation: From Massachusetts to the World," a permanent exhibit at Boston Logan International Airport.

Directly or indirectly, these innovations have fundamentally changed the world. They include the birth control pill, which shook society when it became available to the public; the robots that went into the reactors at Japan's Fukushima installation after the Tōhoku earthquake and tsunami; and

America's first public school, which showed the path to a better future for the nation. To showcase all five hundred would have resulted in too cursory a review for each; these carefully chosen fifty illustrate the breadth and depth of the Boston area's contributions.

Punching above Its Weight

Boston has defied epochal changes in the global economy, repeatedly emerging in a new and more vigorous form, no matter the challenges it has faced. Its phoenix-like existence reveals much about what it takes for a city or region to weather storms, adapt to change, and build a resilient living culture that is strong and enduring.

As the innovation thought leader Scott Kirsner puts it, although not a particularly large city, Boston consistently "punches above its weight." By some measures, the city's metropolitan region is the sixth largest economically in the United States and ranks twelfth globally, with annual economic activity of more than $438 billion. And Massachusetts is one of the top three states in terms of the annual value of venture capital investment. Boston's sports teams have been among the most competitive in the country, giving it a visibility out of proportion to its size. And its professional sporting legacy is deep; with its opponent, Pittsburgh, the city hosted (and won) the first baseball World Series in 1903, and its professional basketball, football, and hockey teams have long traditions of national success.

Today, one of the most noticeable things about Boston—one often remarked upon by outsiders—is its educational and research enterprise. The metropolitan area attracts 250,000 students to the region annually at its sixty-plus colleges and universities. Among them are some of the most famous in the world, especially Harvard, the oldest college in the United States, and the Massachusetts Institute of Technology (MIT). However, these two heavyweights are joined by almost another dozen major research universities, including Northeastern University, Boston College, Boston University, Wellesley College, Tufts University, and the University of Massachusetts Boston.

This concentration of academic talent, which includes researchers and educators from around the world, has helped sustain the city as a home to technology-oriented companies. In the past generation, the region has become a center for biotechnology, while its long-standing leadership in medical research has grown even more substantial. Indeed, the Milken Institute, an economic think tank in Santa Monica, California, rates Boston as the most important life sciences region in the country. A recent study by Luís Bettencourt, of the Santa Fe Institute in New Mexico, found that over the past twenty-five years, more scientific papers in the top three science journals were authored by researchers in the Greater Boston area than by scientists from any other city worldwide.

But the innovation economy doesn't just stem from the institutions of higher learning. Large research universities attract talented people, but they don't, by themselves, make a city innovative on a world-class level. So much of the culture of innovation exists and thrives not just inside university walls but outside them as well. And a world-class research university does not by itself transform a mid-sized city into one filled with related innovations and hundreds of successful companies. New Haven, Connecticut, home to Yale University, or the town of Princeton, New Jersey, did not morph into what Boston has become. Nor did Cambridge and Oxford in England. It is not only the major university that makes the critical difference. Something else is also needed.

Boston has a habit of continuing to redefine and reinvent itself. Whether through new products and services or new industries, the region has repeatedly given itself a new and promising future, even when the immediate outlook was bleak. To be sure, the great innovations of one era—such as the clipper ship, the innovative machines that wove cotton and wool, or cotton

A State of Nobel Prize Winners

On the level of ideas alone, Boston and Massachusetts have had incredible success. More than 170 Nobel Prize winners have conducted their significant work in Massachusetts, or they have worked in the state for significant portions of their professional career. In fact, if Massachusetts were an independent nation, it would rank second in the world to the United States as a whole for the total number of Nobel Prizes, ahead of the United Kingdom and Germany.

and rubber firehoses—led to the abandoned factories or shipyards of the next. And the same may very well happen with today's biotech and robotics industries. Do we know what will become of the infrastructure that supports today's modern-era inventions such as the mutual fund, the robotics industry, the labs creating a cure for Gaucher's disease, or Akamai Technologies' development of online streaming? Economic and political forces beyond the control of local areas are continually picking winners and losers. Yet through four major economic upheavals in history, Boston has been driven back to success by four waves of innovation.

The Core of the Story: Boston's Innovation Drivers

At their most basic stage, innovations are often a by-product of people interacting, bouncing ideas off one other, trying to solve one challenge and then another, seeking to be the first to develop a product and then scale it and bring it to market. Cities, with their diverse individuals, interests, and subcultures, are the perfect setting for innovations and their creators. Most critically, cities like Boston help bring together the five drivers of lasting, prominent innovation. These drivers may, at first, sound so natural that they are not worth mentioning. But their effects are profound. Why has the Boston region continued to develop new ideas and then produced innovations that have kept the economy growing, replacing those same industries and products with new industries, when so many cities have not?

A group of project teams, formed by the Boston History & Innovation Collaborative (BHIC), set out over the course of more than a dozen years to focus on this question and find the answer. With a research team drawn from twelve universities and a working board of directors, this group examined in detail a representative subset of the nearly five hundred Greater Boston innovations drawn from technology, medicine, finance, education, and society during nearly four hundred years of the region's history. The trustees were diverse and included a county sheriff; a woman whose family came from West Africa; the head of a national insurance company; the electrical engineer who became MIT's president; independent scholars; Pulitzer Prize–winning historians; biotech leaders; and entrepreneurs who were of many races and ethnicities, some American-born and others more recent immigrants.

These teams looked at why each of these innovations occurred in the Boston region rather than anywhere else. Why did Alexander Graham Bell come to Boston to work on his telephone research rather than settle in Canada, London, or Scotland—all places he had worked before? The teams spoke with hundreds of scholars and laypeople and asked the hard questions. Why is this region so innovative and what drives that innovation? With few natural assets, did the area's early settlers scramble to find industries that could help their towns survive? Was the early drive for public education in Massachusetts the main factor? Or did the networking among individuals in a smaller city provide the foundation? Or was it because Boston was the second largest seaport for immigration into the United States?

The teams boiled their answers down to twenty-four different drivers and then focused on those that proved most essential. They tested the level of significance of each driver for each innovation. At each stage, the teams involved hundreds of people to be sure that they were looking widely for answers to why Boston was home to such innovation. From this exhaustive study, the teams found that five of the drivers stood out, drivers that, over most of Massachusetts's history, explain why the innovation had happened in the Boston area, and why at that particular time.

The Five Drivers

1. Strong Entrepreneurship

Each innovation was linked to a strong entrepreneur who persisted, no matter the obstacles, in bringing the innovative idea over the finish line. By *entrepreneur*, we mean a person who energized others, who created something revolutionary that changed that moment in time for the betterment of people or the world. Such entrepreneurs or pioneering teams were people who creatively solved problems, invented ways to fulfill common demands, and pushed with everything they had—and more—to make a new solution work.

A good example is Charles Goodyear, who, in 1844, figured out the formula to make rubber that would last in all types of weather. He went to debtors' prison for years, working while incarcerated over a small burner to create a durable rubber.

Then there was Dr. Judah Folkman, who successfully invented a means for restricting blood supply to cancer cells, a huge breakthrough in reducing cancer for many. His persistence as a scientific entrepreneur is legendary. Folkman developed his theory in the 1970s, though his ideas were denounced for decades as unprovable and he was viewed as an outlier among researchers.

A used clothing salesman and the freed son of African American slaves, David Walker, is famous for *David Walker's Appeal to the Colored Citizens of the World* of 1829, still considered one of the most important antislavery documents, which mobilized many to take action. While living in Boston, Walker advocated for the total abolition of slavery and the recognition of black people as free Americans, who belonged by right in the United States, rather than people who should be returned to Africa, a policy widely advocated by some antislavery activists. He played a key role in energizing the white abolitionist movement.

The role of the entrepreneur is essential in bringing together the other drivers and achieving ultimate success. But the interaction of the innovator with the other four drivers is crucial. Entrepreneurship cannot exist on its own and relies mightily on the others. Together they create (choose your metaphor) a secret sauce, a perfect storm, a moment ripe for a lasting and momentous innovation.

2. Local Networking

Many large cities are built and sustained around one major industry, but Greater Boston has usually had three or more major industries thriving at one time. Its industrial diversity has consistently given innovators the opportunity to capitalize on an impressive range of information, ideas, and services that they learn about and use by turning to others in different industries.

Inventing and innovating happens when pursuit-driven entrepreneurs get feedback on their initial ideas. Many people believe inventing is solitary, not collaborative. But Boston's innovations have emerged from an environment where people look for solutions to puzzling problems while listening to others and their feedback. Discussing ideas with others and listening to the others' comments can be just as important as the individual inventor's work and research. This driver has been vital to almost all of the innovations studied, and networking is crucial in most of the fifty stories in *Boston Made*.

In the early 1900s, King Gillette had been working on the idea of a disposable safety razor in his apartment in Boston's Back Bay neighborhood. He sought the help of a colleague to find someone who could manufacture steel blades supple enough not to skin the shaver, something which all the other metallurgists Gillette spoke with at MIT and other places said was impossible. Through his circle of acquaintances, he met with an MIT engineer, William Nickerson; within a month, Nickerson had made the famous double-edged razor a mechanical reality. And within three years, the razor had begun to reshape shaving habits internationally. The name Gillette was synonymous with razors, and the inventor had made a small fortune.

When inventor William Otis was working on prototypes for a steam-powered excavator, he had input from all kinds of metalworkers and blacksmiths at a nearby manufacturer, which made all the difference. Learning about their work and listening to their feedback helped him create a successful design that led to orders from nearby railroads hoping to speed up their construction projects.

3. Local Funding

The availability of local, as opposed to national, funding has been critical in Boston's innovative past and continues to be so today. Local funding provides resources to invest in risky ventures. In the seventeenth century, it was the wealthy who helped to fund the first public school. At its inception in 1635, Boston Latin, the first public school in the United States, received much of its funding from the new town's wealthiest individuals and a lesser sum from a public town tax. In the 1800s, thriving merchants, stymied by

President Jefferson's embargo on trade due to the Napoleonic Wars, used their idle money to fund America's first fully integrated textile plants just outside Boston.

Later, in the mid-1840s, Enoch Train, who owned a fleet of commercial vessels, recruited Donald McKay, a talented shipbuilder from Newburyport, Massachusetts. McKay had earned a reputation in New York City building sailing ships. Train saw that McKay needed a large shipyard of his own to invent and build the best ships, and he helped McKay set up such a yard on the East Boston waterfront. The result of their collaboration was the clipper ship. They were the fastest ships under sail at the time, and McKay became the master of the type.

This sort of local funding was first institutionalized by several Boston banks, helping local entrepreneurs forge new enterprises. In the first half of the twentieth century, when the old Boston banks had become more cautious about investing in innovations, one wealthy businessman and professor launched what we now call venture capital. It was the first modern form of funding for innovative start-up companies, and since that launch in 1946, local venture capital has played a key role in innovation and reinvention of the region's economy.

4. Local Demand

Having early *local* adopters, both businesses and individuals, who would purchase these new innovations, especially when the fledgling companies didn't have the funds to venture into larger markets, made a huge difference. Local demand not only allowed opportunities and innovations to be tested and improved, it also provided enough cash flow to stay in business while the innovators prepared to meet the demand of a national or international market.

Take, for example, the first underground subway in the Western Hemisphere and the third in the world. It was 1897 and the subway/trolley system (later named the Massachusetts Bay Transportation Authority [MBTA], also known as the T) was constructed in Boston to reduce an hour-long daily stall of street trolley traffic along a main thoroughfare in the city. Before it was built, many locals said they wouldn't be willing to travel underground, and there were concerns that the entire endeavor would be a commercial flop. Instead, after the underground subway was completed, crowds jammed the new transportation system and kept doing so. Boston's subway soon became a benchmark for others around the nation and around the world.

Software innovator Dan Bricklin developed the electronic spreadsheet, the first so-called killer app software to become widely popular in the United States. Working in the late 1970s with his professors at Harvard Business School, where he was enrolled in an MBA program, Bricklin was able to test ideas and quickly generate sales through a local organization. He sold the first VisiCalc software packages through the Boston Computer Society, made up of six hundred early computer-savvy members, including himself. It was an organization that allowed members of the fledgling software industry to connect and discuss the latest breakthroughs in hardware and software. It was only after this early acceptance that Bricklin made direct contact with Apple's Steve Jobs, and the software was able to take off on the national level.

5. Global Demand

The greatest successes occur when an innovation responds to market demands that are global, not just local. In an unusual number of cases, Boston inventors or entrepreneurs, paired with merchants and marketers, have been particularly apt at reaching seemingly unlikely international markets.

Take the story of salt cod. Historically, the waters off Nova Scotia, Massachusetts, and Maine were teeming with codfish. In an effort to turn around a crash of Boston's farm economy in the 1630s, Governor John Winthrop and one of his sons, who had moved to a Caribbean island to start a sugar plantation in the 1640s, figured out a use for cod skins, which could be preserved by salting. This process made an inexpensive food, rich in protein, for the enslaved workers. Boston merchants entered a new global market selling the salted codfish skins in the West Indies and returning with sugar, which was made into rum in Boston. The trade made Boston boom for the next century, although it unfortunately strengthened the slave system of the Caribbean islands.

Another example of an innovation with a global reach comes from Fidelity Investments' owner and CEO Ned Johnson, who realized that his firm's mutual funds would be even more appealing if investors could write checks against their fund balances. Tested first with local account holders in Boston, the concept's reception was so positive that it was quickly marketed globally by the company. It contributed significantly to the loyalty of its customers and the company's rapid growth in subsequent years.

Bump and Connect

Of the five factors that were most critical to innovation in Boston—strong entrepreneurship, local networking, local funding, local demand, and, ultimately, global demand—the first four involve proximity, a shared feature we like to call "bump and connect." Bump and connect includes the planned and unplanned meetings and conversations that occur among people, for example, an entrepreneur with other entrepreneurs, experts, financiers, researchers, or simply friends and acquaintances. Each can play a role in helping to supply missing ideas, inspirations, and connections to resources, people, and/or money, or simply supply feedback or a critique. All of these bump-and-connect elements are like a well-tested recipe in which each ingredient, or factor, affects the others, and can help turn research and inspiration into something that ends up changing the world.

These informal meetings used to happen (and sometimes still do) in coffeehouses, in pubs, and at office desks and tables, or when the innovators were walking the same streets or attending the same evening lectures or social events. Today, telephone and internet connections have become part of the phenomenon. Bump and connect sparks collaboration across and within industries or communities and increases the rate at which ideas advance and become actions, products, cures for diseases, or new ways of thinking.

At its center is the notion that the isolated individual does not drive innovation, although that individual is very important. Rather, the way in which individuals can discuss ideas is what is important. When people are able to meet, whether serendipitously or by design, they are in a position to share ideas and

Introduction

Boston has been a center of innovation for four centuries. Newer innovations here mimic earlier ones. The Bunker Hill Monument, commemorating those who died fighting the British in the Battle of Bunker Hill, is mirrored in the innovative design of the Leonard P. Zakim Bunker Hill Memorial Bridge. Innovation building on innovation.

services more easily, which can lead to new products, industries, and social movements. Boston is not the only place where bump and connect has been important. But its combination of strong local drivers and its extensive bump and connect have positioned it uniquely for innovation success. Boston is probably not alone in being driven by most of the top five factors. And it would be instructive to know what drives innovations in other major cities or regions that have not been studied for innovation as closely as Boston has.

The idea of bump and connect has bubbled up in many innovation analyses. Popular innovation writer Steven Johnson, in his bestselling book *Where Good Ideas Come From: The Natural History of Innovation*, analyzed hundreds of innovations and found that the "coffeehouse" model, which emphasizes dynamic effects of interaction, is vastly superior to the "microscope" theory of how big breakthroughs develop. Johnson looked at how and why two famous British scientists at Cambridge University—James Watson and Frances Crick, one with a physics background and the other in chemistry—successfully conceptualized and revealed the twin upward spiral structure, or the double helix of DNA, and won the Nobel Prize for their work. Together, they spent considerable time between labs ambling along country roads discussing how DNA might be structured and eventually achieved an insight that proved correct. At the same time that Watson and Crick were walking, talking, and meeting at the local pub, Rosalind Franklin, a biologist in London who believed that microscope-based research would allow her to find out the then-secret structure of DNA, ended up coming up short and missing the prize. Johnson documents dozens of similar dynamics in his book.

To add more weight to the notion that the interaction of inventors and innovators with others is at the core of the process, Martin Reuff conducted a California study examining conditions under which one group of MBA entrepreneurs make large breakthroughs, while other groups make much less. Reuff found that those entrepreneurs who reached out to people from different backgrounds, or with whom they had "weak ties" (people who wouldn't ordinarily travel in their professional or social circles), were *three times more innovative* than the group that didn't reach out to such a diverse set of people. The power of reaching out to those who are different is what makes bump and connect so valuable.

The takeaway is that you are more apt to get some of your best ideas when talking with a work colleague from a different division of your large company, a relative who works in a different field, or a stranger who doesn't have a similar background. These connections are vital to overcoming the most challenging barriers to innovation while helping you to invent something new or different. As some of your tough problems lie dormant in the back of your mind, the interruption and interaction with someone or even something different can contribute a new perspective in solving problems. Most innovations are an amalgam of different pieces of existing ideas or technology: new views stimulate and lead to ultimate success.

As you read the fifty innovation stories in Boston Made, drawn from such a wide variety of fields, you will see how important interaction, as well as the other four drivers, are in almost every case. The concept of bump and connect is not just the interconnecting; networking, while critical, is not enough to make the innovation happen. The interweaving of the other drivers, including local funding, local demand, and the never-say-die entrepreneur, taken together, make Boston so unusually successful with innovation.

A Core Example: Kendall Square in Cambridge

Over the past thirty-five years, Kendall Square in Cambridge, Massachusetts, once a manufacturing area, has been transformed into the epicenter of a global biotech industry. Using vast new knowledge about genes and DNA, companies based in Kendall Square have made major breakthroughs in treating a wide variety of illnesses, including hepatitis C, multiple sclerosis (MS), and cystic fibrosis. In 1990, there were just two biotech companies in Massachusetts, both in Kendall Square; now there are more than a hundred in that one neighborhood, with nearly one thousand in the state. Together, they employ more than seventy thousand people statewide, helping Massachusetts earn its spot as the top employer for biopharma in the country and around the world. And even in the digital age, biotech executives in Cambridge tout the benefits of proximity in innovation.

The five driving forces are all there in Kendall Square, of course: motivated entrepreneurs, many from nearby universities; networking and information exchange over coffee or on a random walk in the neighborhood; local funding from venture funds, many based right in Kendall Square; local demand, especially from five large research and teaching hospitals within a radius of a few miles; and global demand, with international pharmaceutical companies always on the watch for the next important breakthrough. Perhaps it is unsurprising, then, that in 2006, when Novartis, the world's largest pharmaceutical firm, decided to move its global research center from its home in Switzerland to a better-connected site, it scoured the world, from China to California to Burlington, Massachusetts, and in the end selected Kendall Square because of this proximity.

Diversity in Innovation

Innovative breakthroughs are made by people of various backgrounds and ethnicities from all over the world. Twenty of the top fifty innovations in our study had significant involvement from women, people of color, and immigrants. Today, one of every five patents originating in Massachusetts is filed by a first-generation immigrant. When you listen to coworkers, colleagues, or acquaintances from other nations and from different racial, ethnic, cultural, and educational backgrounds, your good ideas can become even better and completely new ideas may be spawned.

Massachusetts didn't begin as a diverse community, but its trading relationships introduced global influences and eventually led to greater human diversity in the region. The first enslaved Africans were brought to Boston just eight years after its founding, and by the 1760s, nearly one in twelve Bostonians were African or African American, some free but most enslaved. Thanks to language in the 1780 Massachusetts Constitution, the state became the first effectively to abolish slavery and went on to become a hotbed for abolitionism in the nineteenth century.

An enslaved African American named Mumbet, who listened to the words of the new Massachusetts Constitution of 1780 about "equality of all men," was inspired to contact a local lawyer who agreed to bring her suit for freedom to the local court. She won not only her freedom but also damages. The case was a test of the state's eighteen-month-old constitution, and the new state supreme court ruled that the case had no standing due to the "equality of all men" clause in the preamble to the US Constitution. By implication, slavery had already been abolished with approval of the Massachusetts Constitution two years earlier, and Mumbet was free. She chose the name Elizabeth Freeman as her new name. From the 1830s to the 1910s, Boston was seen as one of the two or three most racially tolerant large cities in the country. (It is worth noting that tolerance then is much different than what most would accept as equality at the beginning of the twenty-first century.)

Like many American cities, the Greater Boston area also absorbed waves of immigrants, beginning in the early seventeenth century, and today is among the nation's most diverse communities, with émigrés and their descendants from Europe, Africa, the Near East, South Asia, East Asia, and the Americas. Boston is a major port of entry for immigrants gaining entry to the United States. All of which leads to a diversity of individuals and in turn contributes significantly to the culture of innovation in the region.

Innovation as Reinvention: The Four Waves

Over the course of Boston's history, the city has had four major economic collapses or periods of stagnation, as economist Edward Glaeser first reported. Yet each time the city reinvented itself, largely because of innovative new industries. Many other US cities and regions that were once the darlings of innovation did not recover their cutting edge after economic collapse, including Hartford, Connecticut, for its precision work with interchangeable parts manufacturing; New Bedford, Massachusetts, for whale oil; and Troy, New York, for iron. But for Boston, four waves of innovation followed the collapses, continuing to give life to the region and in some cases spreading nationally or globally.

The First Wave

In the 1630s, as Puritans arrived in the Boston area, the farm economy in this region boomed. As each new wave of English immigrants arrived, the overwhelming demand for farms and cattle drove up the price of both. In 1641, English Puritans and the British parliament they controlled drove Charles I from the throne. A number of leading Puritans in the Massachusetts Bay Colony, including some of its wealthiest, returned to England to fight for the Puritan and parliamentary cause. As some of the wealthiest Puritans sold off their resources, prices dropped dramatically for land and cattle in the Massachusetts Bay Colony, as the region was called. Similar to a bursting housing bubble, this collapse left many financially strapped, even ruined, forcing a number to declare personal bankruptcy, including the governor of the Massachusetts Bay Colony, John Winthrop.

Winthrop led an effort to save the Massachusetts Bay Colony by building the infrastructure for a new shipbuilding industry and mercantile economy. The tall, ancient trees of Massachusetts were a key product; they made the best masts for the British navy and new Massachusetts-built ships. Winthrop also worked with one of his sons to bring the latest ironworking technology from Britain and thus built the colony's first ironworks, which fueled the success of the shipbuilding industry. Until that time, the British Crown wouldn't allow iron tools or pieces used for shipbuilding to originate in the colonies. The colonists also fished the abundance of cod in the region and built a substantial and lucrative trade with the sugar-producing islands of the Caribbean. Within eighty years, Boston was the third-largest port in the British Empire, and the largest town in British North America. Nearly a third of Boston's population was involved in the shipping trades. Innovations in trade and shipbuilding, like innovations in Silicon Valley in the past forty years, made the Greater Boston area a boom town.

The Second Wave

In the 1750s and 1760s, the economy of Boston was again stagnating. Following the costly Seven Years' War, the British government began taxing the American colonies, and the people of Boston were especially resentful. Boston, which twenty years earlier had been the largest port in what is now the United States, was declining; unemployment was rising. Boston merchants, artisans, and dockworkers protested, most notably through their destruction of heavily taxed tea shipments at the infamous Boston Tea Party. The British retaliated and shut down the Port of Boston. The British Army occupation of Boston from 1768 to 1776, the battles of Lexington and Concord of 1775, and later the battle to end the British occupation of Boston effectively liberated Massachusetts when the British were forced to evacuate and turn their focus to New York (it would be another five years before all thirteen colonies were liberated from British rule). By the time Massachusetts had effectively won its independence in March 1776, nearly 85 percent of the population had moved, died of smallpox, or left to join the new Continental Army. Boston was struggling.

In 1783, when the peace treaty with Great Britain ending the Revolutionary War was signed, Boston's port reopened, although it was still prohibited from trading with the British sugar islands. Boston's economy began to return very slowly. The merchant John Hancock was elected the first governor of the newly independent Massachusetts Bay Colony. In an effort to develop new trade routes, Hancock and others established the first state-chartered bank. This bank largely funded a trip to China to establish a new trade route, which Hancock and Sam Adams, another innovative leader driving the new state of Massachusetts, created when they received a letter from John Ledyard, a New England college dropout who had visited China. That trip brought otter pelts, highly desired in China, from the uncharted Northwest, now the states of Oregon and Washington, across the Pacific Ocean to be exchanged for China's most desirable products.

These merchant-traders made the region remarkably wealthy and developed ever-more competitive ship types. Boston sea captains took a tremendous risk trading in the Northwest with the Native Americans along its coast, and succeeded. The risk of sending a ship halfway around the world for nearly two years, based on a single letter, is a great example of one of the drivers: global demand. And it could not have happened without the innovation of the first chartered bank.

The Third Wave

The dominance of Boston sea captains in shipping was ended by the Panic of 1857, a nationwide financial crisis, combined with a quick shift from sailing to coal-fired steamships. But the collapse didn't last long. The Civil War ultimately led to a booming economy in Massachusetts: selling blankets, uniforms, and shovels to the army, as well as rifles and revolvers, all products manufactured in Massachusetts.

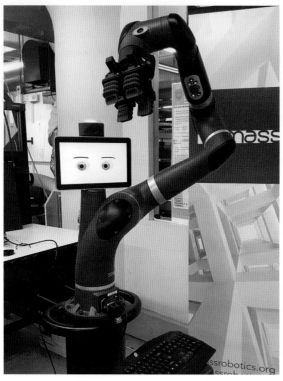

A cluster of Massachusetts robotics companies, including iRobot, have developed a wide variety of machines in the late twentieth and early twenty-first centuries. This robot, built by Sawyer, is on display at the MassRobotics facility.

The region not only survived this turn of fortune but grew wealthier. Traders became investors and managers of new and growing manufacturing enterprises and railroads, turning a nascent scattering of industrial experiments into a wholesale industrial revolution of textiles and shoes that powered the region into the twentieth century. Boston played the key role in the invention of the telephone, for example, as well as the early electricity industry. In manufacturing, Boston, along with cities just south (including Brockton) and north (including Lynn), innovated in the shoe-manufacturing machinery.

The textile industry—both cotton and wool—made the city a center of manufacturing for national and international markets. Boston's market for wool, in what is now called the Seaport District, led the world during World War I. The conversion of the horse-drawn streetcar industry to electric trolley and then to the first subway was also a major innovative breakthrough. Eastern Massachusetts, which had been a hub for carriage manufacturing and for Albert Pope's bicycle craze in the late nineteenth century, became the national center of the new auto industry before the locus of the new industry moved to Detroit in the early 1900s.

The Fourth Wave

The fourth wave, or "The Long Stagnation," as it is called, was really a long, slow decline of major industries and a failure to replace them with new ones. Economic growth slowed from the 1920s through the 1970s. Leading the decline were the key industries of shoemaking and textiles, made worse by the impact of the Great Depression. Massachusetts companies couldn't compete with the manufacture of clothing either, and these industries moved to the southern United States or to other parts of the world where labor was less expensive. Boston's port gradually atrophied.

In the 1940s, there were really two economies: the old manufacturing economy that continued to decline, and a second, smaller but expanding economy based on technology, hospitals, and universities and focused on innovation. World War II provided an impetus to innovations in electronics. Raytheon paved the way as a defense company, with its contributions to the development of radar and tube miniaturization, and there was corresponding growth in universities and hospitals. After the war, there was a new highway system that supported the creation of suburbs surrounding cities in what had been farmland. The growth of the family car and the suburbs was part of the story.

Starting in the 1950s, computer, telecommunications, and other technological companies grew significantly. Innovation drove these sectors along with research hospitals and universities. The pace of economic growth after World War II and since was helped by an expanding finance sector and by federal funding connected to defense industries and research hospitals.

Always Innovating

The ability of Boston to recover from stagnation or collapse and then to thrive as a trailblazing incubator is based in no small part on the diversity of its innovative industries, which have sustained the region through several downturns. Ingrained in its culture is the drive to invent repeatedly and to generate new products and ideas that serve the city, the region, and the world in new and exciting ways. Each of the recoveries from the economic downturns was led by new, innovative industries—the new wave. The city offers a great blend of people, possibilities, resources, and consumers: the right people to develop ideas, others to fund those ideas, and the innovators to drive the ideas until they succeed. The sheer mass and breadth of innovations coming from Boston and Massachusetts are significant.

Today, the Boston region's research universities are a meaningful supporting driver in most fields of innovation. They bring together huge numbers of intellectually curious people who are working as much on campus as they are off. They are driving innovation all the time. But that alone doesn't explain why this region has been so innovative over four centuries. As one top corporate executive said recently in explaining why he was moving the corporate headquarters of his century-old firm to Boston, it is because of the "thick layer of innovation." In his view much more than the research universities make the difference between Boston and other university hubs.

The production and headquartering of telephone systems, as well as other promising industries such as frozen foods, minicomputers, mainframes, and the modern defense industry, have all been part of the replacement of the old textile and shoe manufacturing industries with new, lighter industry. Shoes and textiles first moved to the South before shifting after World War II to Italy, Spain, Brazil, and other nations. Steel and metalworking, nurtured in the Boston area when the colonies and the republic were young, moved closer to the resource-rich areas at the heart of the country. More recently, information technology (IT) moved to Silicon Valley. Yet each time, Boston bounces back.

The following fifty innovations illustrate moments in time, eras of collaboration, and years of determination to make products, create solutions, heal humans around the world, capitalize on needs and desires, and entertain. Each innovation has changed Boston, the nation, or the world for the better. This group of fifty, taken from nearly five hundred examples of innovation in the Boston area, illustrates what it means to become a global center of innovation, and to maintain this status over centuries, through waves of new industries.

BUMP AND CONNECT

2010s Kendall Square
WORLD BIOTECH CENTER

Of critical importance to the world of innovation is what we have called bump and connect. The proximity of academic researchers, entrepreneurs, funders, and other representatives of the five drivers of innovation to each other is an essential part of what makes a region a great place to conceive and develop new ideas and bring them to fruition.

The growth and development of the biotech industry in Kendall Square, in Cambridge, Massachusetts, a community adjacent to Boston, is one of the clearest and best examples of how and why eastern Massachusetts has produced so much innovation over time. We learned in the Introduction why some of the biggest research-oriented pharmaceutical companies have decided to locate in Kendall Square or one of the other areas close to Harvard Medical School and four of the world's leading hospitals.

Studies have shown that the best innovation research happens when researchers are located less than thirty minutes from each other. Kendall Square is a mile and a half from Boston's Massachusetts General Hospital (MGH), often named as the world's number-one research hospital. Massachusetts Institute of Technology (MIT) is just four blocks away. The Longwood medical and academic area, which includes four main research hospitals and a number of leading research labs, is less than four miles away. Harvard University is two stops by subway. Five of the country's leading venture capital firms are also within five miles. And finally, less than six miles away, Boston Logan International Airport, with direct flights to sixty nations, connects researchers and pharmaceutical organizations globally. In addition to this rich proximity of resources, which helped the Massachusetts biotech industry grow by 28 percent over the last decade, the Massachusetts Life Sciences Act of 2007 put $1 billion toward support for the growth of the industry.

This success story began in the 1970s. Some brilliant scientific sparks that might have sputtered and faded in other cities came together with other key drivers of innovation to create something remarkable. Two professors helped spark that early development: Walter Gilbert from Harvard and Phillip Sharp of MIT. Both were Nobel Prize winners: Gilbert for his method of decoding DNA, and Sharp for his discovery of split genes. With other biologists, they helped form a new kind of pharmaceutical company, the Switzerland-based Biogen, in which each of the researchers was a leader in a new field of science called genetic engineering. Biogen moved to Cambridge in the early 1980s: its goal was to use research on human genes to search for cures for diseases that were seen as incurable.

Genzyme (now Sanofi Genzyme) and Biogen were the pioneers. But not for long. From two small companies in the 1980s, the pharmaceutical sector grew to over five hundred companies and nearly seventy-five thousand jobs thirty years later. It is also the story of new innovations in pharmaceuticals and medical treatments improving the lives of millions and saving countless lives.

It is not easy being an innovator. In 1976, four years before Biogen's launch, an intense political debate arose around the new field of recombinant DNA, which was the basis for the newfound ability to blend genetic material from more than one source and grow it inside a living organism. The research sounded like science fiction, and it scared some Cambridge residents. What if the work on genes led to living beings, perhaps dangerous beings, that could escape from the laboratory? Alfred Vellucci, mayor of Cambridge at the time and a populist who liked to suggest that some Harvard research could hurt average Cambridge families, led the opposition. But it wasn't just politicians; another Nobel Prize winner, George Wald, advocated for putting genetic labs in remote locations, away from population centers, rather than in a tightly packed, old city like Cambridge.

The mayor introduced a resolution to place a moratorium on the establishment of genetic labs in Cambridge for two years until more was known. "Is it true that in the history of science, mistakes have been known to happen?" Vellucci asked rhetorically. "Do scientists ever exercise poor judgment? Do they ever have accidents?" Applause broke out in the packed hearing room. Banning genetic laboratories became a big issue for Boston-area television news and a hot-button local political issue.

But the attention worked in favor of the new industry. The hearing led to a vote to set up a commission to report to the city council, and the study concluded in 1981 with a recommendation not to ban the genetic labs. It also proposed safety regulations. Indeed, because of the political battle by Vellucci and others, Cambridge became the first city in the world to establish safety regulations for genetic research. The city's initial opposition is a good example of how innovation can often be highly controversial.

The clarity and workability of the resulting regulations helped Biogen as it built the first commercial genetic labs in Kendall Square. In 1980, the US Supreme Court ruled that work on genetics in living things could be patented, a decision that contributed to the establishment of a legal foundation for this new industry.

However, one important piece was not yet in place: finance. In the 1980s, the National Organization on Rare Diseases (NORD), led by relatives of people suffering from rare diseases, lobbied on behalf of developing legislation that would encourage start-ups focused on rare and deadly diseases. Three leaders in Congress led the fight, most notably Massachusetts senator Edward Kennedy. Kennedy was approached by NORD, universities, and biologists to create this legislation. After years of lobbying and legislative work, the Orphan Drug Act was passed in 1983 to allow companies that developed pharmaceuticals and devices for diseases affecting fewer than 200,000 in the United States to apply to receive research funding and tax credits in de-

Located adjacent to MIT, close to Harvard and numerous teaching hospitals, Cambridge's Kendall Square has proven a fertile place to launch and grow new enterprises. Recently it has blossomed as a global center for biotech and life science. In the foreground is the square, with the Charles River, downtown Boston, and Logan Airport beyond.

veloping their drugs. The legislation, which needed to get approval for funding by Congress every five years, passed with support that Senator Kennedy was able to secure from allies in the Republican Party. This funding provided the basis for the early progress enjoyed by the biotech industry. The industry was still nascent. There was no biotech industry outside Massachusetts apart from a small California company called Genentech.

But Massachusetts proved to be a particularly fertile area, again, because of its early regulatory clarifications. With funding from the Orphan Drug Act, the National Institutes of Health (NIH) went to several Tufts University School of Medicine researchers to see if they would start work on one of the rare diseases. These researchers started Genzyme, which moved into Kendall Square in 1990. By now, venture capital firms were taking notice. The earliest venture capital firm to focus on biotech was T. A. Associates, led by Peter Brooke, a former student of venture capital pioneer Georges Doriot. Genzyme's success in 1991 with the drug Ceredase showed that a healthy profit could be made from biotech innovations. Another critical institution was Kendall Square's Cambridge Innovation Center, where hundreds of start-ups, sixteen venture capital firms, and a weekly Venture Café help innovators and funders find each other.

As life sciences consultant Harry Glorikian has put it, "Being in proximity to others makes a big difference in drug discovery. When you work at the lab bench, with the soft science and the whiteboarding you do in biology, being close by shortens the time and increases the probability of success."

Massachusetts biotech is in a very strong position. Along with California, it leads the nation in this important, innovative industry. Eighteen of the top twenty global biotech companies now have facilities in Greater Boston, most of them in Kendall Square. This number is unmatched by any other city worldwide.

CREATING A BETTER WORLD

1781

From Promise to Reality

THE FIRST STATE TO END SLAVERY

Any time while I was a slave, if one minute's freedom had been offered to me, and I had been told that I must die at the end of that minute, I would have taken it just to stand one minute on God's earth a free woman.

— Elizabeth Freeman

In 1775, as the American Revolution began, the enslavement of Africans and African Americans was legal in all thirteen American colonies. Before the war ended, Massachusetts would make history by becoming the first state effectively to abolish slavery. The cases of two brave African Americans—a female house slave from western Massachusetts, Elizabeth Freeman, and an enslaved male farmhand, Quock Walker, from central Massachusetts—were critical to this historic shift. Both had been brutally beaten by their masters. Independently, both sued for their freedom through jury trials. And in both cases all-white local juries voted them free. The reason cited by the juries: the one-year old Massachusetts state constitution, which declared "all men free and equal." The state's new supreme judicial court upheld the juries' decisions.

The female house slave was known as Mumbet. She was born in 1744 on the property of a wealthy Hudson River Valley Dutch farmer, and at age seven she was given as part of a wedding gift to the farmer's daughter and her new husband, John Ashley, a leading revolutionary patriot from Sheffield, Massachusetts. In 1780, Mrs. Ashley was trying to beat Mumbet's younger sister with a heated shovel; Mumbet shielded the girl and received a deep wound in her arm. She left the scar uncovered in public as evidence of her harsh treatment by her mistress.

Mumbet was intelligent as well as proud. Although illiterate, she had heard a reading of the new Massachusetts Constitution, written by John Adams and ratified in 1780, either in her owner's kitchen or in front of the house in Sheffield. Inspired by the reading and recent beating, when Mumbet heard the phrase "all men are born free and equal" she searched out John Sedgewick, an abolitionist-minded local lawyer, and asked him to take her case. Within a year, John Ashley was ordered "to release unlawfully obtained property." Ashley offered Mumbet a paying job, but she refused, going to work instead for Sedgewick. She worked for his family for decades, changing her name along the way to Elizabeth Freeman and purchasing her own home.

The male farmhand, Quock Walker, was born a slave in Worcester County and worked on the farm of his owner, James Caldwell. Caldwell and his wife had both promised

Quock his freedom, but after Caldwell's death in 1764, his wife married Norman Jennison, who assumed ownership of Quock. No word about freedom was mentioned again.

Quock ran away from Jennison in 1781 and went to live in the house of James Caldwell's brother. Jennison went after him, retrieved him from Caldwell's home, and beat him terribly. Quock pursued the continued enslavement with Caldwell in court, arguing, like Mumbet, that under the new Massachusetts constitution, he was free and equal. The jury ruled in his favor, and the Massachusetts Supreme Court later ruled that the jury had correctly interpreted the state constitution's wording.

Other Northern states followed the example of Massachusetts and abolished slavery, some quickly, others much later. New York, for example, passed a statute abolishing slavery in 1799, but its implementation was very gradual, and final emancipation did not happen until 1827. Of course, it would take nearly a century for the United States to abolish slavery once and for all with the passing of the Thirteenth Amendment in 1865.

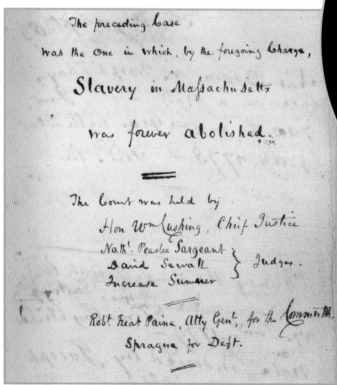

Born a slave, Mumbet played a major role in making Massachusetts the first state to abolish slavery.

Mumbet's bravery and the leadership of Chief Justice William Cushing, whose writing is captured in this court document, effectively abolishing slavery in the state of Massachusetts.

1829
Abolition Redefined
ALLOWING FREED SLAVES TO STAY IN THE UNITED STATES

In spite of widespread belief that it was sinful, slavery was such an important part of the early US economy, and so entrenched in US culture, that the number of Africans and African Americans enslaved in the United States actually continued to grow in the early 1800s, in spite of restrictions in the new US Constitution. Part of the challenge of defeating slavery was that opponents were not necessarily supportive of immediate abolition, and most did not believe in racial equality.

Quakers, who opposed slavery, as well as some free-born African Americans and freed slaves founded the American Colonization Society (ACS) in Philadelphia in 1816 to support sending freed African Americans back to West Africa. A number of free African Americans opposed this notion of recolonization, asserting that they had been born in the United States and belonged here. In some cases, their own families had lived and toiled in the United States for generations, for as long or longer than many white Americans.

Those who opposed the ACS felt it profoundly unfair that their role and contribution to the nation was to be rewarded with banishment. They saw the push to remove free African Americans from the United States as racist and may have noticed that many supporters of the ACS were themselves slaveholders, eager to remove free African Americans from the country. Nevertheless, the ACS moved ahead, and in the early 1820s established an initial colony of about four thousand on the so-called Pepper Coast of Africa, a colony that eventually became the nation of Liberia. For the decade and a half after the ACS was founded, there was no organized movement to oppose its colonialist solution.

In 1829, a free-born North Carolinian African American named David Walker, offended by the ACS and its views, published *David Walker's Appeal to the Colored Citizens of the World* (*Walker's Appeal*), a religiously and emotionally driven pamphlet advocating that freed slaves stay in the United States. Walker wanted all means necessary used to abolish slavery immediately and to recognize former slaves as full-fledged citizens rather than candidates for the questionable idea of repatriation—a revolutionary, indeed innovative concept, for

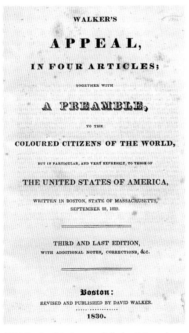

Widely circulated, especially in states where slavery remained legal, *Walker's Appeal* denounced those who would send freed slaves back to Africa to form new colonies. Walker helped shape the growing abolitionist movement to permit former slaves to stay in the US as free members of society.

that time and place. Walker, a clothing merchant, visited many northern cities, and settled in Boston in 1825. Although slavery was long illegal in Massachusetts, many Boston whites were hostile toward blacks, and racially motivated violence was on the rise in the 1820s. However, Walker also found a city where, more than any other city he had visited, more blacks owned their own businesses and homes. Before his untimely death in 1830, his innovative approach to the "problem" of slavery, which the nation's founders had failed to address, had an important impact in two ways.

First, he succeeded in getting *Walker's Appeal* distributed covertly in the South. To distribute the publication widely, Walker had copies delivered surreptitiously by giving them to African Americans and sympathetic whites traveling to other cities, and he used his connections to Boston's shipping industry to distribute his pamphlet to readers in the North and the South. In the South, *Walker's Appeal* stirred up quite a reaction. The state of Georgia offered $5,000 for Walker's arrest, and the mayor of Savannah asked Boston's mayor to prohibit the pamphlet's distribution in the South.

Influenced in part by David Walker, William Lloyd Garrison's *The Liberator* continued calls for abolition, encouraging reformers and enraging slaveholders through the end of the Civil War.

Second, Walker connected with the Boston reformer William Lloyd Garrison. Although he wanted an end to slavery, Garrison still advocated recolonization, and Walker convinced him that immediate abolition was the best and only solution. Walker's argument helped to inspire Garrison to start a religious-based, interracial abolitionist movement, arguing that freed slaves were *Americans* and therefore should be allowed to stay in the United States when freed. Garrison's subsequent meeting with local African American leaders at the African Meeting House led him to found an abolitionist organization that reached out to both blacks and whites. Garrison also started the *Liberator* in 1830, a newspaper initially made possible by donations and subscriptions from Boston's African American community. Counteracting the racist sentiments of many white abolitionists, Garrison's approach was possible because he chose to attend conventions for blacks and enlist their financial support.

Walker and Garrison diverged on the role of violence: Walker encouraged slaves to rise against their masters in armed revolts; Garrison condemned any such rebelliousness. Eventually, the *Liberator* became *the* newspaper of the abolitionist movement in the United States, expanding, and expanding again, until it was nationally famous and influential in the 1850s.

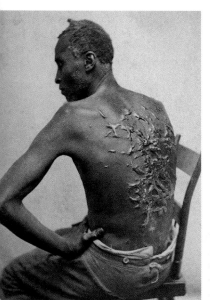

This photograph of a brutally scarred slave, taken in 1863, vividly illustrates the violence that was used to maintain the slave system.

1830s

Dreamers and Doers

AMERICA'S TRANSCENDENTALIST MOVEMENT

After independence, the United States quickly established a new form of government and a way of life that was more democratic than that of England or the European continent. American literary culture remained focused, however, on British models, British philosophical and literary movements, and other Old World sources. The novels, histories, poems, and plays that people read in London were usually what became popular in the United States, and even the literary content produced by US authors remained British in form and feel.

In the 1830s, a group of Massachusetts writers and thinkers from a new generation developed the first US-born philosophical and literary movement: they were called the transcendentalists. The transcendentalists looked beyond British and European intellectual models; rejected much established Christian religious belief, particularly that of their Puritan forebears; and believed instead that all living things were bound together, that humans were essentially good, and that insight was more powerful than experience as a source of knowledge.

This combination of ideas was a major innovation in US culture and had profound social implications. Transcendentalists' forward-thinking attitudes and ideas about morals and issues led them to support the abolition of slavery and a more expansive role for women, although when the Transcendental Club first met in Boston in 1836, only men were present. The opening of the first railroads allowed more mobility within the Greater Boston area, and in addition to Boston, the towns of Concord and Cambridge, and later the communal Brook Farm in Newton and West Roxbury, played important roles. While many think of the transcendentalists as individuals working and writing alone, it was their network—and their frequent visiting and discussions—that led them to a new, distinctly US literary culture.

Henry David Thoreau, today perhaps the most famous of the transcendentalists, was a friend of Ralph Waldo Emerson and an early advocate of civil disobedience as a tool for social change.

In an effort to recapture connectivity to nature and oneself, the transcendentalists sought to go beyond the cold calculations of rationalists, who had come to rely for social meaning on technology and industry. Educators, clergy, schoolteachers, and poets came together to debate and form opinions about the connectivity of life. Although their search was sometimes described as nostalgia for a simpler life, this group was collaboratively seeking a unique structure in which the spiritual growth that connection with nature, individual democracy, and other US institutions afforded. This spirit tied disparate writers and thinkers together, from the most conservative to the most radical.

Today, Henry David Thoreau is perhaps the best known of the transcendentalists. His book *Walden* is both a celebration of his connection to nature—an account of living in a cabin next to his beloved Walden Pond—and of civil disobedience, or passive resistance, a philosophical stance that inspired Gandhi, Martin Luther King Jr., Nelson Mandela, and the Vietnam antiwar movement.

Transcendentalism's most distinctive champion was Ralph Waldo Emerson, a former Unitarian minister who trained at Boston Latin and Harvard College and lived in Concord, not far from Walden Pond. Emerson's speeches and essays, including "The American Scholar" and "Nature," defined why and how the new culture of transcendentalism broke so decisively from the dependence on British literary thought and explored the new perspective so well that he became the leading voice of intellectual culture in the United States.

Emerson's good friend Margaret Fuller overcame much adversity to become a groundbreaking teacher and then the leading female writer, columnist, and editor of her generation. She was the first full-time American female book reviewer in journalism, and her book *Woman in the Nineteenth Century* is considered the first major American feminist work. Fuller died in a shipwreck in 1850, and her memoirs, edited by Emerson and published two years later, was one of the best-selling books of the nineteenth century.

While transcendentalism was relatively short-lived and not without its critics (Nathaniel Hawthorne among them), it had an important influence on later American thinking and literature. It even helped shape the philosophy of some leading thinkers in India later in the 1800s.

The transcendentalists embraced new ideas, some of them quite radical, and included a number of prominent women. Margaret Fuller was an editor, writer, educator, and advocate for women's rights.

Henry David Thoreau found Walden Pond, on the outskirts of Concord, Massachusetts, to be a spot for inspiration and contemplation, leading to publication of the eponymous *Walden* in 1854.

1863

Fighting for Respect

THE FIRST AFRICAN AMERICAN REGIMENT TO FIGHT A MAJOR BATTLE

From the end of the American Revolution to the Civil War, free African Americans were strongly discouraged from joining the US Army, and those who did join were, with few exceptions, not allowed weapons, a policy justified by the popular myth that African American troops would break and run in the heat of battle. When the Civil War began, a group of abolitionists, including Massachusetts governor John A. Andrew, argued that the war was about African American freedom from slavery; as such, it should include African Americans fighting on the Union side. For two years this group lobbied President Lincoln to authorize an African American fighting regiment, but their efforts to create this social innovation did not bear fruit until 1863, when Lincoln issued the Emancipation Proclamation, declaring that African American slaves in territory controlled by the Confederate forces were free.

Lincoln then agreed to let Massachusetts establish an African American unit: the 54th Massachusetts Infantry Regiment. To lead it, Governor Andrew appointed a twenty-five-year-old white colonel, Robert Gould Shaw, who recruited more than a thousand troops and trained them in Readville, a neighborhood south of Boston. Shaw, as well as Boston's free blacks, had to fight each step of the way in their efforts to have the African American unit taken seriously. They feared that the regiment would be made into a support unit for white Army regiments: handling logistics, building earthworks, repairing railroad tracks, and performing other secondary tasks for which escaped slaves had previously been employed by the Union army.

Trained with rifles to fight in battle, the 54th Regiment proudly marched up Beacon Hill and past the Massachusetts State House

The Massachusetts 54th was recruited with the help of Boston's black community and abolitionists and included two sons of Frederick Douglass, who had escaped enslavement and become a prominent national voice for the cause.

Private William J. Netson was one of a number of regiment members to be photographed. He survived the war and lived until 1912.

The Second Battle of Fort Wagner was one of the bloodiest moments of the Civil War. Famous at the time, and made the subject of a popular movie, *Glory*, in 1989, the 54th fought with great bravery to help Union forces tighten their blockade of Charleston, South Carolina.

on its way to board ships that would transport its members to South Carolina. But once in the South, they were, as they had feared, kept from battle for six weeks. They were also paid substantially less than white soldiers, and the regiment members refused to take any pay unless it was equal to other soldiers. Meanwhile, Shaw argued vociferously to let his men fight in battle. Finally, on July 18, 1863, Shaw was given the order for the 54th Regiment to be the lead unit in an assault on Fort Wagner, which guarded the entrance to Charleston harbor. Charleston was the city that had effectively sparked the Civil War with its assault on Fort Sumter.

The novelty of a predominantly black unit leading this assault attracted a considerable number of northern newspaper reporters, who witnessed a disciplined advance as the 54th Regiment marched across the beach toward the fort and attempted to clamber over the ramparts. In the face of long odds and heavy casualties, the newspapers reported, the 54th fought with valor and tenacity, repeatedly reaching the fort walls only to be shot down. The regiment's casualties were very heavy, and Shaw was among the dead. When the assault was finally abandoned, the 54th Regiment had only 354 men left who were capable of fighting.

The political implications of the 54th Regiment's role in this battle were extensive. Several other Northern states created armed units of African American soldiers. Although the Union model of having white officers command segregated African American troops was maintained by the US Army until 1948, the display of courage by the 54th Regiment that July 18 was a turning point. The myth of African American troops breaking and running proved false. A number of other Northern states embraced setting up units with African American soldiers armed with rifles. One black soldier of the 54th had retrieved the unit's flag, the vital rallying point in the confusion of battle, after the designated flag-bearer had fallen. He continued to guide his fellow soldiers forward as he was wounded multiple times and was later awarded the Congressional Medal of Honor, the first such award for an African American. Although it took another year before Lincoln and Congress agreed to pay African American soldiers the same as white soldiers, black soldiers were granted equal pay for the whole time they served. By the end of the Civil War, 200,000 African Americans had fought for the Union.

2003 I Do

GUARANTEEING THE RIGHT OF LESBIANS AND GAYS TO MARRY

To many in favor of gay marriage, Massachusetts was viewed as a state highly unlikely to pass legislation approving marriage equality, largely because of its large Catholic population. The opposition of the Catholic Church to gay marriage rights was a very powerful influence in a state with so many Catholic citizens. The issue was brought to the fore in the early 2000s, however, when investigative reporting revealed multiple stories of clerical sexual abuse. The resulting scandal undermined the Catholic Church's influence and helped set the stage for the Massachusetts State Supreme Judicial Court's 2003 ruling in *Goodridge v. Dept. of Public Health*.

The *Goodridge* ruling was the culmination of a grassroots effort that spanned two decades and involved a wide range of individuals, nonprofits, law firms, religious groups, and finally the courts and state legislature. This collaborative effort helped kick off other national and state movements, which transformed the whole country on the issue of same-sex marriage. The legality of same-sex marriage was established at the federal level and thus in all fifty states following a US Supreme Court ruling on June 26, 2015.

As early as 1971, Boston's gay pride movement began to grow noticeably, and with each passing year, more and more lesbian, gay, bisexual, transgender, and queer or questioning (LGBTQ) individuals felt comfortable coming out in Boston and rallying to the cause of equality. Organizations and coalitions set up to fight for this cause included the Massachusetts Gay and Lesbian Political Caucus

Massachusetts chief justice Margaret Marshall (left), who wrote the state's supreme court's opinion, and GLAD attorney Mary Bonauto, who made the case to the court for marriage equality for all in Massachusetts.

City Hall, Cambridge, Massachusetts, May 17, 2004: The freedom to marry takes effect in Massachusetts, and same-sex couples begin marrying for the first time ever in the United States of America.

(MGLPC), founded in 1973 to lobby for gay civil rights in the state legislature, and Gay and Lesbian Advocates and Defenders (GLAD), founded five years later to protect gay citizens from unjust discrimination by law enforcement. The Massachusetts branch of the American Civil Liberties Union (ACLU) also became active during this time in promoting and protecting the rights of gays.

As more gay couples began to adopt, family rights and the need for marriage equality became a pressing issue. Gay couples demanded protections for their children, and gay men, more aware of the lack of legal recognition for their relationships as they lost loved ones to AIDS, rallied around the call for marriage equality. In 1993, the Freedom to Marry Coalition of Massachusetts was formed to campaign for same-sex couples to be entitled to the thousand-plus rights and protections associated with marriage.

As these organizations worked to secure equality for the LGBTQ community, it became apparent that a new approach would be necessary to bring about enduring change. When lawyer Mary Bonauto joined GLAD in 1990, she and the rest of the team knew they had to bide their time and gain support among the general population before raising the issue of same-sex marriage as a legal matter. They needed a political climate conducive to a fair trial and worked alongside several organizations and individuals to establish the groundwork for such a case.

Bonauto stressed the necessity of having the public recognize that the case was about their neighbors, colleagues, friends, and family members. She considered carefully which couples should be selected as plaintiffs when GLAD brought the *Goodridge* case to court. On November 18, 2003, the Massachusetts Supreme Judicial Court (SJC) decided that the Massachusetts Constitution "forbids the creation of second-class citizens" and "affirms the dignity and equality of all individuals." Mary Bonauto and Chief Justice Margaret Marshall, who wrote the case opinion, are considered the intellectual leaders of the decision. The Massachusetts SJC also gave the state legislature 180 days to take whatever action was necessary to align itself with the ruling.

During this time serious opposition arose. Governor Mitt Romney, the office of the attorney general, and the state senate chose to interpret the *Goodridge* decision as an authorization of civil unions rather than marriage. Dozens of LGBTQ and non-LGBTQ civil rights organizations, along with ninety leading constitutional law professors from across the country, joined GLAD and appealed to the court in a brief opposing civil unions as an alternative to marriage rights. In an advisory opinion, the Massachusetts SJC responded that "the history of our nation has demonstrated that separate is seldom, if ever, equal."

During this critical time, a coalition of marriage-equality supporters formed MassEquality to protect the *Goodridge* decision from an anti-same-sex marriage amendment to the state constitution, which would have invalidated the 2003 state supreme court ruling. State legislators went back and forth on the issue, with increasing support for marriage equality after each round of discussion, until June 2007, when the last attempt to put this constitutional referendum on the statewide ballot was finally defeated in the legislature. In the months leading up to the legislature's vote, MassEquality organized a strong grassroots political campaign.

As it had with the abolitionist movement of the eighteenth and nineteenth centuries, Boston and Massachusetts led the fight for marriage equality. The Greater Boston area's diverse population, including people like Mary Bonauto, were willing to challenge conventional wisdom and create the innovative organizations listed above for a successful legal, political, and social movement.

ENHANCING PUBLIC LIFE

1634

For the People

AMERICA'S FIRST URBAN GREEN SPACE

We expect every city to feature shared pastoral spaces, public parks, where citizens can mingle and seek refuge from urban life amid greenery and water. We take such parks for granted now, but America's first public park, established on Boston Common in 1634, was a true innovation. No other British North American colony had chosen to pay for urban land for a common space (though such spaces were common in Great Britain).

To begin, the first park felt more rural than urban. Four years after the Puritan settlement of Boston, inhabitants voluntarily taxed themselves to buy forty-eight acres for grazing cattle. This space was so popular that twelve years later, in 1646, the town had to cap the number of cows at seventy, and over the century that followed, the pasture would evolve into a model of what a democratic urban park could be. It was widely copied.

History and need shaped and sometimes delayed this evolution. To try to control what it viewed as Boston's unruly population, which had widely opposed the Stamp Act and

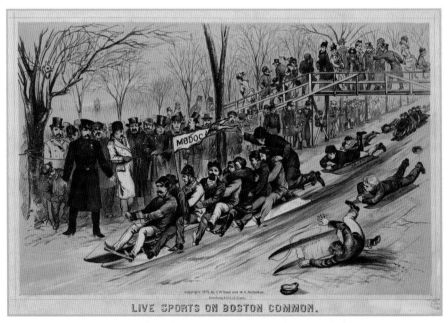

Boston Common still has Frog Pond—used for skating in the winter and as a spray pool in the summer. As shown in this 1875 print, in the past, the common hosted more boisterous activities.

Boston Common has witnessed history and remains a meeting ground for all near the center of the city.

other acts of Parliament, the British Army occupied Boston in 1770, housing hundreds of soldiers in tents on Boston Common. During that winter, schoolboys used the steep hill near what is now the State House to swoop down on sleds into the encampment and taunt the occupiers. These soldiers cut down huge numbers of the trees for firewood and embarked in their boats at the Back Bay end of the Boston Common to cross the Charles River and fight the patriots in several battles of the American Revolution. By the time the British were forced out by General George Washington and his new Continental Army in 1776, large parts of the park were barren of trees and grass. It took years to bring the space back.

By 1830, Boston had grown into a true city and ended the grazing on the Boston Common, which now spread below the new State House on top of Beacon Hill. The common's Frog Pond became a place for boys, girls, and families to skate. Fountains and monuments were built. And within a few years, an iron fence enclosed the common and its five internal promenades, including the Tremont Mall, an imitation of St. James's Park in London. With the broad walkway in place, Boston Common could claim to be the world's first public urban park, preceding the establishment of the earliest such parks in England—Derby Arboretum (1840), Peel Park in Salford (1846), and Birkenhead Park (1847)—often described as the first urban parks in the world.

From celebrating British surrender in 1781 to Martin Luther King's civil rights rallies of the 1960s to marches by gay marriage advocates, Boston Common has become a sacred civic space in the middle of downtown Boston and the center of Boston's life. This unique park, an innovative Massachusetts treasure, has helped inspire land preservation and conservation across the nation.

1635

Democratic Education

AMERICA'S FIRST PUBLIC SCHOOL

The Massachusetts Bay Colony at Boston, established in 1630, was not the first European settlement in what would become the United States. Jamestown, Plimouth, and New Amsterdam preceded it. But it was the first to establish a public school: in 1635. It was open to all boys regardless of class, and it operated under the direction of the government, it was supported by taxes, and it was not connected to any church. New Amsterdam (later the city of New York) didn't establish a public school until the 1680s, fifty-seven years after the settlement was founded. What was it about Boston that drove such an innovation just a few short years after its founding?

Some would argue that the importance Puritan settlers placed on having children learn from reading books (the basis of their religion was reading the Bible) was the catalyst. And the religious driver was significant: when John Winthrop preached to his fellow Boston-bound Puritans in 1630, he directed them to pursue a high standard of conduct and purposefulness, in accordance with their interpretation of Biblical directive. Others argue that the idea of the public school and the importance attached to it arrived with immigrants from East Anglia, where the original Boston is located and which had a higher level of schooling than the other areas where most of the English Puritans came from.

Regardless of the exact impetus, with only a few thousand residents clinging tenuously to a peninsula, Boston needed more than tradition or religion to give birth to the first school in the colonies. It took an entrepreneur of sorts, John Cotton, one of the key thought leaders in the new colony, to make it a reality. Cotton had attended the Free Grammar School in Boston, England, and urged fellow citizens of the new town of Boston to authorize the use of public funds to create a new school.

Operating in borrowed quarters initially, thanks to both public and private support, the Boston Latin School soon graduated to a series of larger buildings. Historians differ as to whether this was the first or second such building.

Philemon Pormort, recently arrived from England, was appointed as the first schoolmaster, and classes were conducted initially in his own home. The study of Latin and Greek were central to the new school's curriculum. Latin in particular was deemed essential for any learned profession or for a life of the intellect. The language of the Roman Empire had persisted through the Renaissance as the language of the church, scholarship, and diplomacy, and the most esteemed Roman and Greek authors of antiquity continued to be studied in schools and universities well into the twentieth century. In addition to their ongoing cultural influence, the discipline and commitment needed to master these subjects also commended them to educators.

Although the school was originally supported largely by donations from the town's wealthiest individuals, such as Henry Vane, John Winthrop, and John Cotton himself, the vote for its establishment in 1635 was tied to a decision by Bostonians to tax themselves to pay for it. Underscoring the importance attached to education in the new colony, the legislature voted in 1642 to mandate that all parents instruct their children in reading and writing. Establishment of public schools in any town of more than fifty families for that purpose was mandated a few years later. Still, there was so much support beyond public funds alone that by 1671, private contributions to Boston Latin School, as it became known, including a number from Boston, England, led to the establishment of a board of trustees to manage the funds.

Over its lengthy history Boston Latin was not only the usual destiny for the city's elite, it was also a nursemaid to greatness for individuals of every social and economic level. Innumerable prominent Americans, from Benjamin Franklin, Samuel Adams, and John Hancock in the colonial era to Joseph Kennedy, George Santayana, journalist John King, media titan Sumner Redstone, Parkland, Florida, mayor Christine Hunschofsky, and Nation of Islam leader Louis Farrakhan got their start at the school. (Benjamin Franklin is Boston Latin School's most famous dropout!)

Nearly four hundred years after its founding, Boston Latin School still operates as a prestigious public school, with admission based on examination. Since 1972, the school has included both boys and girls from grades 7 through 12, and it is currently ranked thirty-three in the top one hundred US high schools according to a report by *U.S. News & World Report*. The school's Latin motto, *Sumus Primi*, for "We are first," refers to the school's early origin as well as to the academic performance of its students.

Boston Latin School remains a nationally ranked secondary school and has a diverse component of male and female students.

1636

Founding a Colony and a College

HARVARD'S BIRTH ON THE EDGE OF WILDERNESS

You might think that starting a college would be the last thing on the minds of settlers arriving on a new continent who had recently experienced a period of near-starvation. Yet that is just what happened in the Massachusetts Bay Colony, where Puritan colonists established the first college, a scant six years after settling in Boston in 1630. Until the twentieth century, no other political entity—colony or nation—started a college so soon after settlement or independence. The colony of New Amsterdam (the colonial name for New York), settled in 1623, did not establish its first college (Columbia) until 125 years later. Virginia, first settled at Jamestown in 1607, didn't establish William and Mary until 1688. What was going on in the Massachusetts Bay Colony to make this educational innovation happen so quickly?

The Puritans sought to create a new community based on what they saw as a purification of the Church of England and its society. Harvard College, located in the town of Cambridge, across the Charles River from Boston, was set up to educate ministers of the Puritan religion. But the speed of this decision by the settlers may have had its roots in internal dissension. For Puritans, there was a good deal of conflict over whether one could be one of the "elect," as defined by their theology, through good deeds alone. Some within the community argued that one didn't need a minister to be part of God's elect. A spiritual leader (although not a minister), Anne Hutchinson, who had conducted discussions with other women and a few men, was expelled from the colony soon after its establishment for arguing this radical perspective. The colony's legislature voted to found the college several days after Hutchinson's 1636 conviction for heresy, and it's possible that it was eager to reinforce orthodoxy in the wake of her departure.

Some innovation scholars have suggested that a culture of intellectual argument might have been a foundation for innovation in Greater Boston. The religious and political debates that filled Puritan public life and the colony's General Court (the original name for the legislature) led to a culture that prized the pursuit of critical thinking, both in the new college and in the town of Boston across the river. These intense public debates set norms in Massachusetts that are reflected in the innovation culture of the region, and which are qualitatively different from the cultures of rival early colonies.

Did the early culture of theological discussion continue into the more secular centuries to come, leading to new ideas and new inventions long after the Puritan environment was gone? We can't say yes definitively, but it is an intriguing thought for this book, which seeks to encourage debate on the spirit of innovation and inquiry.

Harvard College as it looked in the eighteenth century.

Harvard took its name from John Harvard, a local Puritan minister who, when he died in 1638, left his library and half of his modest fortune to the new college. The town of Cambridge also adopted its name in 1638, in tribute to Cambridge University near the English town of Boston, where John Harvard and other leaders of the new colony had studied for the ministry.

Over time, Harvard grew from a college for ministers to a full-fledged university, adding a medical school in 1782; law school in 1817; and gradually schools of engineering, architecture and design, education, public health, business, government, and others. The college also grew secular in outlook, although Harvard Divinity School continues the tradition of religious education.

From its founding in 1636 through the 1830s, the primary source of financial support for the college was from the colony, then state, of Massachusetts. The only college in the American colonies for fifty-seven years, it was a virtual national institution throughout that period, drawing students from the Atlantic seaboard. As colleges were gradually founded in the South and Mid-Atlantic states in the eighteenth century, Harvard became more provincial, educating primarily males residing in the Northeast. It returned to national

Linked to a nearly four-hundred-year educational continuum as the first American college, Harvard University grew tremendously through the twentieth century, moving well beyond its original regional focus to become a global institution.

prominence in the nineteenth and twentieth centuries under the leadership of a series of presidents who helped set the direction for US higher education right up until today.

Harvard's national and global prominence continued to grow, as it has become a force for education, business, health, government, the arts, science, and engineering. More Nobel Prize winners have researched and/or taught at Harvard than at any other university in the world. Even so, in the 2010s, some critics have asserted that the university has not been innovative enough, leading Harvard to work harder to develop innovative medical and IT capabilities that could lead to commercial industries rivaling what MIT has done in Kendall Square, Cambridge.

1775–1776

Birth of a Nation

THE FIRST INDEPENDENT STATE IN THE UNITED STATES

The creation of the state of Massachusetts—the first independent state in the United States—was itself an innovative event in US history. A mere eleven months after Paul Revere's ride, a colony was transformed, the British Army was forced to evacuate, and the first state came into being.

Innovation starts with examining an existing way of handling a problem and thinking of different ways to handle it. As entrepreneurial leaders emerge and reach out to a diverse group of people, networks develop, sometimes in opposition to the existing way of doing things. Some militants stand their ground against the status quo, conflicts develop among those who want to stick with the current model, and new notions (in the Revolutionary era, first liberty, then independence) emerge from the intellectual and military battles.

In this sense, the events that led to an independent United States were truly innovative, beginning with strong opposition in port cities such as Boston to the British Parliament's Stamp Act in 1765. In Boston, masses rallied in opposition to the Stamp Act under the Liberty Tree (a famous elm tree and gathering point near Boston Common); other towns in the colonies rallied in opposition, too, and eventually the act was repealed. "No taxation without representation" was an old concept given an innovative meaning.

In Massachusetts, Committees of Correspondence were developed to keep each town's leaders in contact with each other in a network of thinkers about the new situation much like the formal and informal networks from which almost all innovation develops. After the Boston Massacre in 1770, when four thousand British soldiers occupied Boston and killed protestors at the State House, the British closed the port of Boston to all shipping, depriving the community of its principal source of livelihood. The traditional militia in the towns of Massachusetts were reinvigorated and began to train, choosing their own officers and creating a group that could be available for battle on short notice: the Minutemen.

How did the British Army respond to this colonial innovation? Fearing the growing militancy, it sent seven hundred soldiers to seize guns and powder stored inland, at Concord. Revere then made his famous ride to alert its guardians. The British won a first skirmish at Lexington Green, where the Minutemen were shot down. But word had spread, and a few hours later the local Minutemen—now with a huge mobilization from towns around Boston—turned back the British at Concord and nearly turned the British retreat into a rout. By the end of the day, the Patriot forces numbered nearly fifteen thousand, coming from as far away as southern New Hampshire. Facing such overwhelming numbers, the British forces were bottled up in the

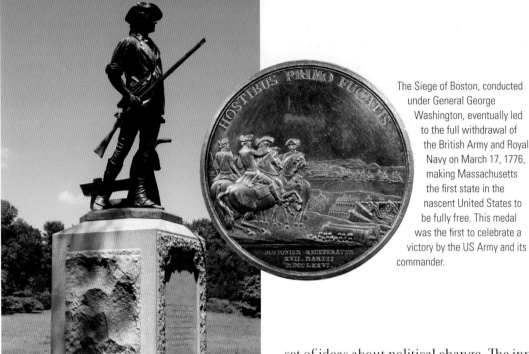

The Minute Man statue by Daniel Chester French at the Old North Bridge, in Concord, Massachusetts, commemorates the success of the farmer "citizen-soldiers" called to arms from across the region to repel the British Army's attempt to capture arms and ammunition from the colonists, on the first day of the American Revolution, April 19, 1775.

The Siege of Boston, conducted under General George Washington, eventually led to the full withdrawal of the British Army and Royal Navy on March 17, 1776, making Massachusetts the first state in the nascent United States to be fully free. This medal was the first to celebrate a victory by the US Army and its commander.

town of Boston, which at the time had only one narrow neck of land connecting it to the mainland. The volunteer militia army laid siege, determined to force Britain to withdraw, from April 19 through the following March. During that time, they fought another, larger battle at Bunker Hill. This unanticipated series of events shows the high level of support and organization that had formed around a set of ideas about political change. The innovation was popular enough that tens of thousands were willing to risk their lives for it.

Recognizing how an army had emerged overnight from its citizenry, the Continental Congress, which was meeting in Philadelphia, authorized the Minutemen encamped in Cambridge to become the army of the Continental Congress and chose George Washington, one of the most seasoned military men in the colonies, to go to Cambridge and take command on July 3, 1775. A year before its formal declaration in Philadelphia, the US Army officially began when Washington took command of a citizen's army with very egalitarian values—itself a major innovation in a former British colony. Six months later, the army flew the first American flag—with thirteen red and white stripes.

A colony in rebellion must have a government, so a state legislature, composed of the sympathetic members of the colonial legislature, met outside Boston as the de facto government of the new state of Massachusetts. Literal and figurative battle lines were forming throughout the other twelve colonies, which followed Massachusetts's lead by creating new institutions that embraced the sentiments of the people. As with many innovations, something new, an independent "state," was beginning to emerge.

In Boston, there was little hope of achieving a decisive victory while the guns of the Royal Navy protected loyalists and the British Army. But another innovation, engineered by a former Boston bookseller, Henry Knox, gave the colonists a critical edge. Moving scores of powerful cannons taken from the British outpost at Fort Ticonderoga, in what is now upstate New York, the colonists made a secret winter's journey over weeks, establishing the weapons on Dorchester Heights, just south of Boston. The power of the new bastion, capable of raining down destruction on the British fleet and the town itself, convinced the British Army and many of its Tory supporters to depart Boston rather than face the American cannon. (The date of their departure, March 17, 1776, is still celebrated in Massachusetts as Evacuation Day.) Fourteen weeks later, the Continental Congress met in Philadelphia. With the victory in Massachusetts in mind and with high hopes for the future, it declared national independence.

On that night of March 17, Massachusetts became the first liberated colony, now a state led by an elected state legislature and independent of Great Britain. It became the cornerstone of what would become the United States of America. General Washington now moved the Continental Army from the Boston area to New York to begin the fight to expand the liberated territory.

Boston paid a high price for its independence. Its shipping and trade economy were prohibited from trading with Britain and its colonies. This led to the worst economic collapse in the city's four-hundred-year history. Almost 85 percent of the population had left the town or died, either in battle or in a smallpox epidemic. It took a decade to recover economically and socially.

1780

Ideals Made Practical

THE CREATION OF AN AMERICAN SYSTEM OF GOVERNING

John Adams, already an inventive leader in Massachusetts before the American Revolution, played a vital role as diplomat in Europe during the war and was the principal author of the new Massachusetts Constitution, still in force today, 250 years later.

Massachusetts was the first of the original thirteen American colonies to become independent of Great Britain. For several years at the beginning of the American Revolution, the Massachusetts state legislature operated as the de facto government. Its power could easily be challenged, however, and pressure soon mounted to develop a new form of government. What were the new state's underlying values? How could the values that emerged in the first few years of the American Revolution be used to create the best system for this new government?

The same legitimacy question was faced by all thirteen colonies. In 1778, the Massachusetts legislature developed a new constitution, which was rejected by the voters in town meetings. After this failed start, the legislature called on each town the following year to elect men to represent them in a constitutional convention, a political gathering unprecedented in the American colonies or around the world. (New Hampshire had used an election to choose delegates, and its constitution had been voted down by the people of the state.)

To develop a blueprint that could be discussed and debated at the new convention, in the fall of 1779 the state turned to three of its revolutionary leaders: John Adams, Samuel Adams, and James Bowdoin. They in turn chose John Adams to propose a new system of government. A student of the classics, a Massachusetts lawyer, and a political leader, Adams had studied other classical government systems. As the US government's ambassador to the French Court and to the Dutch Republic, he had learned firsthand about European forms of government.

Adams first developed a document of general ideas and liberties based on a core set (a bill) of rights and on the belief that "all men are born free and equal, and have certain natural, essential, and unalienable rights." Second, he developed a framework that emphasized checks and balances to discourage tyranny, or domination by only one group. This framework included three co-equal branches of government (legislative, judicial, and executive), separation of powers for each branch from the other two, an elected executive (governor) chosen by popular vote, and a two-chamber representative legislature.

What Adams developed was largely approved by Samuel Adams and James Bowdoin, followed by the full convention, and finally voted on by people at town meetings. In a first, the constitution was approved by referendum, at the time a very innovative method of choosing a new form of government. Patriot Boston merchant and Continental Congress president John Hancock was elected the first governor under the new constitution. Perhaps more important, within eighteen months two county juries used the new constitution's wording relative to all men being born free and equal to free an enslaved woman and an enslaved man, effectively ending slavery in Massachusetts (an innovation in itself for the thirteen new states).

The constitutional system developed by the state of Massachusetts influenced other states as they developed constitutions. Eight years later, when a national constitutional convention was held in Philadelphia, the template was no longer controversial. It played a contextual role. Today the Massachusetts Constitution is the oldest full constitution in continuous effect in the world, and many nations have adopted variations on the same basic framework developed two hundred and forty years ago.

Among other things, the Massachusetts Constitution included a three-branch government, each with different powers, and a bill of rights. That and other features of the document were an important reference point for authors of the US Constitution.

1839

Teach the Teachers

AMERICA'S FIRST PUBLIC TEACHERS' COLLEGE

Massachusetts was the first colony or state to have a public school in what became the United States, and Horace Mann was the first secretary of education for the state of Massachusetts. Mann is known as the father of American public education, and as secretary of education, he founded the first statewide system of public education and invented a new approach to raising the standards of local public schools.

Core to Mann's approach was his development in 1839 of the first college to teach teachers, an educational innovation with a feminist perspective: Mann's college was for women, opening up the profession at a time when women teachers were considered an oddity. Mann's wife, Mary Peabody Mann, also a teacher in a school she started, was among the most ardent feminists of the era.

Mann called his college a normal school, a term he took from France, where an *ecole normal superior* educated teachers. Later, these normal schools became state teachers' colleges and today, typically, state universities. The first teachers' college was also the first public women's college in the United States and among the first ten colleges established for women in the country. Its students and graduates were known as normalites.

A lawyer by training, Mann was a political leader and president of the state senate, and it was expected that he would be elected governor. Always interested in new ideas, he had limited schooling in his youth and had never taught. As a farm boy, he attended a town school that had a school "year" lasting just twelve weeks. The governor had pushed Mann hard to take the job of secretary of the Massachusetts State Board of Education; Mann considered it career suicide and initially rejected the offer, but the governor pushed harder and Mann finally accepted the role. High schools were a new innovation at the time, limited to a handful of towns and cities in Massachusetts, and Mann recruited Cyrus Peirce, head of Nantucket High School, to start the normal school. He gave Peirce twelve weeks to create it.

Horace Mann

A commencement speech at Antioch College that Horace Mann gave two months before his early death in 1859 served as a call to graduates. He asked them to embrace his influential worldview: "I beseech you to treasure up in your hearts these my parting words: Be ashamed to die until you have won some victory for humanity."

The college opened with three young women in a school with sixteen books—what the innovation world today calls "Minimum Viable Product," or MVP. It might have seemed an impossible situation, except to Mann and Peirce. Almost no girls were allowed in the few public high schools, and most of the young women had only finished grammar school, so it was difficult to figure out how to predict who most deserved entry to the normal school. Also, no textbooks existed to teach teachers because it had never been done, and at first there was no curriculum for the new school. Mann had decided to make the normal school a boarding college because the young women had to travel long distances. Soon afterwards, Mann also opened two other normal schools for men in other parts of the state.

The first normal school, based in Lexington, Massachusetts, eventually moved to Framingham, and the student body was soon racially integrated. During the latter years of the Civil War and the Reconstruction years that followed, women teachers from Framingham Normal State School were at the center of new schools for freed, usually illiterate African American slaves under the Freedmen's Bureau. Later graduates helped start the first normal school in South America, in Argentina. And in the 1880s, Booker T. Washington, who had started Tuskegee Institute in Alabama, recruited Olivia Davidson, an African American graduate of Framingham Normal State School, to help open the college (eventually one of the best African American colleges in the segregated South).

Mann went on to become a congressman and continued to champion the spread of public education as well as abolitionism. He later served as a university president of Antioch College before his early death in 1859.

Framingham Normal School, the nation's first teachers' college and one of the first women's colleges, was also the first to teach women educators.

MATTERS OF LIFE AND DEATH

1721 An Innovation from Africa

THE FIRST ANNOUNCED INOCULATION IN THE WESTERN WORLD

A smallpox epidemic, carried on a Royal Navy ship from Barbados, raged in 1721 in Boston, the largest town in British North America and the third-largest port in the British Empire. The powerful and influential Boston minister Cotton Mather, who had watched three of his own children nearly die from the disease, urged doctors to begin performing inoculations.

At the time, quarantine—the strict separation of those who had smallpox from those who didn't—was the standard medical practice in the American colonies, as it was in most of Europe. However, Mather had learned of the practice of inoculation from his enslaved African, Onesimus, who had been inoculated as a child in Africa. The medical process then consisted of making a cut in the arm and dropping in a small amount of pus taken from a smallpox sore of someone who was already infected.

Against hostile public opinion, Dr. Zabdiel Boylston agreed to try the technique, initially on his only son and two slaves. For some time afterward, Dr. Boylston was a prisoner on his own farm due to the fury of his neighbors and colleagues. (One of those opponents was a young apprentice named Benjamin Franklin.) Mather also felt the public's wrath—someone threw a lighted grenade into his home.

To test the innovative African practice, Dr. Boylston inoculated a total of 244 citizens. The results could not be denied. For those who did not get inoculation, the death rate was one in twelve; after inoculation, it dropped to one in forty! Two other doctors followed his lead, inoculating people in Cambridge and Roxbury. Successful inoculation trials like these eventually led to the development of vaccination against the disease.

Word of the success of the trials spread: the case was written up in Boston for the Royal Society in London, and even some of the Brit-

> What about Onesimus? His ethnicity was identified by Mather as "Guaramantee"—possibly central Sudan or the coast of Ghana. His medical innovation was not acknowledged at the time, and when Mather later freed him, it was not because he had been instrumental in ending the epidemic but because he was "froward"—what today we would call disobedient.

ish royal family were inoculated within two years, in small part because of the report from Boston. But lost in the success was the fact that the innovation rested directly on the testimony of a man brought to the colonies as a slave. In the early eighteenth century, accepting advice from an African and a slave was taboo, especially for something as fundamental as an innovative medical technique in a situation when many might die. Boston had shown that turning against racial prejudice was more than morally just—it was innovative, in this case solving a frightening public health problem. Boston would display this kind of leadership several times in its history, including when Massachusetts became the first state to make slavery illegal.

Inoculation was almost unknown in the western world when Dr. Zabdiel Boylston, inspired in part by the testimony of an enslaved African, Onesimus, successfully inoculated part of the Boston population against smallpox.

A map of Boston in 1722, the year after the first successful smallpox inoculation.

1846

Modern Surgery Created

THE MIRACLE OF ANESTHESIA

Until the middle of the nineteenth century, surgery was a terrifying and painful experience for the patient. Although alcohol, opium, hypnosis, and a variety of herbs were all used in an attempt to eliminate pain, none worked reliably. Patients remained alert, communicative, and *aware* during even the most invasive surgeries, a situation almost unimaginable today. But an unknown Boston salesman turned dentist, William Morton, building on the research of a network of physicians and scientists, learned about sulfuric ether from a professor. Acting on a faster track than his professor, or his former dental partner Horace Wells, Morton created the first successful anesthetic used during surgery at a hospital.

In the 1840s, Boston was a nascent center of a science education industry that encouraged medical and dental practitioners to be open to innovation. Like many others, including Horace Wells, Morton was experimenting with ways of relieving the tremendous stress of surgery. Wells had carried out experiments with nitrous oxide and had been invited by Dr. John Collins Warren, head of surgery at Massachusetts General Hospital (MGH), to demonstrate. Wells's trial run failed. After additional failed trials, Dr. Warren extended a similar invitation to Morton, who used sulfuric ether.

To perfect his solution, Morton had used Boston's network of experts, including the tradespeople who produced the inhalers, showing how Boston had the medical resources and the openness that any truly innovative culture requires. His experiment, on October 16, 1846, on a young man undergoing surgery to remove a neck tumor, was a success.

It is difficult for us to appreciate the risk that the prominent Dr. Warren was taking. The historian Julie Fenster, in her book *Ether Day*, has noted that when Dr. Warren asked

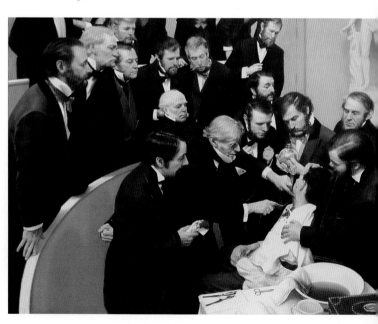

This painting memorializes the first public demonstration, in Boston in 1846, of the use of ether as an anesthetic.

Massachusetts General Hospital, founded in 1811, is the third-oldest general hospital in the United States. The dome at the center of the building is where the first demonstration of ether occurred, and it has been known as the Ether Dome ever since.

Morton to perform the ether demonstration, he "had everything that his profession could offer, in terms of respect and, more than that, of trust . . . many people around the hospital wondered why Dr. Warren, at the age of sixty-eight, would risk that respect and that trust on an unknown entity such as William Morton and a panacea believed to be impossible: a painkiller for use in surgery."

The world reaction to this experiment was swift. Just after the surgery, a ship left Boston for Glasgow with a man who had just learned of the experiment. Upon arrival, the man ran up the hill to Glasgow Hospital with the news, and that same day it was tried successfully there. The breakthrough was quickly communicated to London and Paris. (It is noteworthy that the Philadelphia medical establishment condemned Morton as a quack. A Philadelphia newspaper quipped, "If it came from Boston it must be humbug.")

Morton's claim as sole discoverer was very controversial. In the years following the MGH demonstration, three other men championed the importance of their contributions. Horace Wells claimed the primacy of his work with nitrous oxide. A southern country doctor, Crawford W. Long, claimed to have successfully administered ether to patients three times after 1842, although there had been no outside observers. Boston chemist Charles Jackson claimed to have suggested to Morton that he use ether as an inhalant. These claims led to several lawsuits and much vigorous politicking, but the question of who discovered ether's use as an anesthetic was never definitively resolved. The claims of all parties were so strong that the statue in Boston's Public Garden that celebrates the innovation of using ether in surgery has four sides, one for each claimant (Morton, Wells, Long, and Jackson).

> "What surgeon is there who has not felt, while witnessing the distress of long painful operations, a sinking of the heart, to which no habit could render him insensible! What surgeon has not at these times been inspired with a wish, to find some means of lessening the sufferings he was obliged to inflict!"
>
> —John Collins Warren,
> *Chief of Surgery, Massachusetts General Hospital, during the mid-nineteenth century*

1947

Never Say Never

FINDING A TREATMENT FOR CHILDHOOD LEUKEMIA

In the mid-1940s, the prognosis for leukemia was the same as it had been when the disease was first identified in 1845: death, often painful, usually within weeks of diagnosis. Eighty-five percent of those with childhood leukemia died. But in the June 1948 edition of the *New England Journal of Medicine*, a young pathologist from Children's Hospital in Boston published the results of a new drug tested on sixteen children suffering from leukemia. Ten of the sixteen had achieved remission from the disease.

That young doctor, Sidney Farber, had learned that folic acid stimulates the growth of bone marrow. He believed that if a drug could block folic acid, the production of the abnormal marrow of leukemia could also be stopped. His drugs, which created folic acid antagonists and slowed or stopped tumor growth, became known as *chemotherapy*, a first in cancer treatment.

Farber's findings were met with skepticism by the medical research community. Thinking at the time was that children with cancer were going to die and should be made as comfortable as possible. Farber was meddling in things he should leave alone. After all, how could a young pathologist, an outsider in the medical research community, with little funding, staff, or scientific equipment, make such a discovery? Yet questions and interest poured in from practicing pediatricians. Farber persisted, and today he is acknowledged as the inventor of chemotherapy.

History played a role in this innovation. After World War II, leaders of a charitable organization formed by the entertainment community, the Variety Club of New England, were looking for a local scientist to support, and Farber's name came up. The small outpatient clinic that he had opened in the basement of Children's Hospital attracted the club's attention, and it established the Children's Cancer Research Foundation to fund Farber's efforts. On May 22, 1948, *Truth and Consequences*, a famous national radio show, introduced a young cancer patient at Children's Hospital named Jimmy to the audience. This broadcast began the flow of contributions that funded construction of the Jimmy Fund Building, which later became the Sidney Farber Cancer Institute.

That radio broadcast also turned the young cancer patient into a symbol for all children with cancer. First the Boston Braves, then in 1953, after the Braves had left for Milwaukee, the Boston Red Sox made the Jimmy Fund the charity of choice for the team. At the time, sports teams rarely got involved in such serious issues. But the Jimmy Fund became *the* fund to support in the Greater Boston area.

> Dr. Farber would always say: "In cancer, the child is the father to the man. Progress in cancer research at the clinical level almost always occurs in pediatrics first."

As with so many other Boston innovations, much of the funding came from Boston sources. This movement toward social responsibility changed sports: other teams in major league baseball, and then throughout professional sports, adopted charitable causes to demonstrate their community-mindedness.

In many respects, Farber was ahead of his time. He came up with the idea of what we now call total care. He made sure that all services for the patient and family—clinical care, nutrition, social work, and so on—would be provided in one place, and that *the staff should work as a team* to plan and provide the treatment together. For the rest of his life, Farber led the fight to find a cure for cancer for children and adults. The support of this cause, like the development of chemotherapy and the concept of total care, were innovations spurred by Farber.

Largely because of his pioneering research, five-year cure rates for childhood leukemia reached about 50 percent by 1970, and they reached more than 85 percent today. In 1983, the Cancer Institute was renamed the Dana-Farber Cancer Institute in recognition of Dr. Sidney Farber and philanthropist Charles A. Dana. Today, Dana-Farber, in Boston's Longwood Medical and Academic Area, is globally renowned for using basic and clinical research to improve the treatment of adults and children with cancer.

Dr. Sidney Farber made enormous strides in treating childhood leukemia and laid the foundation for modern chemotherapy. The Dana–Farber Cancer Institute in Boston is named in his honor and that of the principal donor. It continues to contribute to advances in cancer treatment.

1954

The Gift of Life

THE FIRST SUCCESSFUL ORGAN TRANSPLANT

Ron Herrick wasn't thinking in 1954 about making medical history that would one day save thousands of lives. He just wanted to save one person: his brother, Richard, who had been diagnosed with acute nephritis, a kidney disease for which there was no known cure. Ron couldn't stand seeing Richard slowly wasting away, and during one visit blurted out in the hospital room, "Can't I give him one of *my* kidneys?" It was just a cry in the wilderness. The public health doctor caring for Richard dismissed it: there had never been a successful organ transplant. But then the doctor remembered that the men were twins, and he reached out to a surgeon at Peter Bent Brigham Hospital (now Brigham and Women's Hospital), Joseph Murray, who had been conducting research into kidney transplants.

Physically transplanting an organ was possible at this time. But the risk of rejection by implanted organs was incredibly high, so attempts at organ transplantation were last resorts. By transplanting the kidney of one twin to another, Murray sidestepped the problem of organ rejection because rejection had been shown not to be a barrier if the transplant was from one identical twin to another. Murray spent months training himself and his team so that they might attempt the transplant. They had the advantage of a recent invention: a machine that could mimic kidney function, a primitive dialysis machine, so that Ron's kidney function could continue during the procedure. Murray spoke regularly with a wide variety of experts in the field both in and outside Boston hospitals.

There was no DNA analysis in 1954. To ensure that Ron and Richard were identical, Murray followed a number of steps, including having the two fingerprinted at a local police station. The police tipped off the newspapers to the story, and the headline "Brigham Doctors to Try Daring Operation" brought much public attention. One Boston newspaper blasted them for "playing God."

A devout man, Joseph Murray reviewed the ethical questions with members of the clergy. Priests, rabbis, and ministers weren't enthusiastic, but they didn't see it as evil. Boston's Roman Catholic archbishop, Richard Cardinal Cushing, said, "If we think it is acceptable for a man to lay down his life for his fellow man, surely the giving of a kidney would be in the same category." The night before the surgery, Murray prayed with his wife for the transplant that he was about to undertake.

The first successful organ transplant was performed at Boston's Peter Bent Brigham Hospital in 1954.

The transplant was a success, and Richard Herrick would marry a recovery room nurse he met after surgery. Eight years later, the disease returned, the kidney gave out, and Richard died, but today, organ transplants have become almost a regular occurrence, and artificial organs are being developed to replace human organs for patients in need.

Murray went on to perform the world's first successful transplant involving non-identical individuals and was an international leader in the development of immunosuppressive treatments to combat the problem of organ rejection. In 1990, he received the Nobel Prize for his work in transplantation.

1976 Vision Quest
BRINGING THE PRINTED WORD TO LIFE FOR THE BLIND

It had been an impossible dream, achieving direct access to the ordinary printed word for vision-impaired people. But that became the goal of Ray Kurzweil, who capped his childhood interest in computers and inventing with training at MIT as an engineer.

He was halfway to creating a reading machine in the early 1970s when he launched Kurzweil Computer Products, Inc. He then led the development of software that could give a scanner the ability to accurately recognize most printed text, potentially allowing documents or books to be captured in full, editable form. Kurzweil saw that when combined with new charge-coupled-device (CCD) scanners, an improved digital scanning technology developed at Bell Labs, and nascent text-to-speech software, he had all the ingredients needed to create a reading machine.

By January 1976, Kurzweil was able to unveil the finished product with the leaders of the National Federation of the Blind (NFB), which had supported the work by ordering six preproduction models. During a news conference, the Kurzweil Reading Machine was received enthusiastically and later featured prominently on the news broadcast of veteran journalist Walter Cronkite, who used it to convert his printed "sign-off" message into synthetic speech.

The machine was not only seen as revolutionary by the blind community but by everyone as a harbinger of the amazing technologies on the horizon. Later, Kurzweil was able to sell his invention to Xerox. Ultimately, thousands of the machines were built.

Of course, technology continued to advance, allowing smaller and less expensive reading machines to be created. In fact, by 2009, Kurzweil was able to demonstrate his software by turning an ordinary cellphone into a reading machine.

Kurzweil didn't stop inventing. In 1978, he adapted his reading machine technology to commercial purposes, creating a machine for reading documents that was quickly adopted by LexisNexis to scan legal reports and news pages, creating one of the first online database services.

Among his many subsequent activities, Kurzweil, from a chance encounter with blind musician Stevie Wonder, took to heart the latter's lament that keyboard synthesizers of the time couldn't emulate more of the world's diverse musical instruments. Inspired, that same year Kurzweil launched Kurzweil Music Systems and two years later introduced the

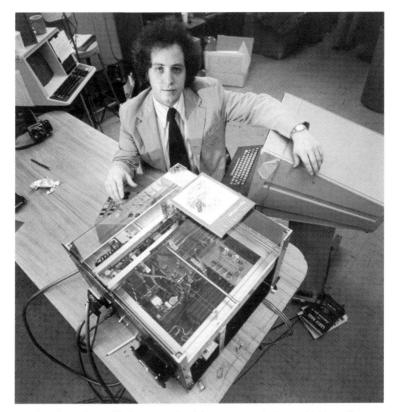

Ray Kurzweil, a prolific inventor and serial entrepreneur, is shown with his reading machine for the blind, hailed as a breakthrough when it debuted in the 1970s.

Kurzweil K250, a keyboard synthesizer that could "play" several different instruments, easily convincing professional musicians they were hearing the real thing. After several more years of development, the company and its inventions were sold to Young Chang, a Korean musical instrument maker.

Kurzweil has been honored with the 1999 National Medal of Technology and Innovation and was inducted into the National Inventors Hall of Fame in 2002. In addition, he received the $500,000 Lemelson–MIT Prize in 2001 for his accomplishments as an inventor.

1996

Science Creating Hope

THE INFANT BIOTECH INDUSTRY'S FIRST BIG BREAKTHROUGH

Multiple sclerosis (MS) is a disease of the central nervous system that disrupts the flow of information between the brain and body. It is tough to predict and often disabling, but many sufferers have been given hope by interferon beta, a drug developed by Biogen. Biogen is a biotech company based in Cambridge, Massachusetts, that conducts research and develops innovative therapies.

Founded in Geneva, Switzerland, in 1978 by an international group of scientists working in the budding field of genetic engineering, Biogen was run by two Cambridge-based scientists, Walter Gilbert of Harvard and Philip Sharp of MIT. It was the world's third biotech company and the first independent biopharmaceutical company.

Gilbert and Sharp both made substantial contributions to the understanding of DNA and later helped with research for the Human Genome Project. At Harvard, Gilbert developed a quick and easy method for reading the chemical composition of DNA, making it easier for scientists to conduct genetic research. Gilbert won a Nobel Prize for this work. Philip Sharp's research allowed scientists to differentiate between "junk" DNA and meaningful genetic material. He also won a Nobel Prize.

Interferon beta was originally identified by Buffalo, New York, researcher Lawrence Jacobs. Jacobs approached Biogen, which in turn formed a research team and used several additional patents to produce the drug, which Biogen called Avonex, and enlist patients in clinical studies. These clinical trials demonstrated positive results, and within six years Avonex won US Food and Drug Administration (FDA) approval.

Intended for patients with the relapsing form of MS, Avonex slows the rate at which the disease destroys nerve tissues while lessening side effects such as vision problems and poor coordination. Scientists speculate that the drug regulates the immune system's response to myelin, the protective outer sheath of nerves. Initial testing of the drug showed that it slowed the progression of nerve damage up to 37 percent over the course of two years and reduced flare-ups by 33 percent. Side effects were negligible, causing a mere 4 percent of patients to discontinue treatment.

More than a half million people have been treated with Avonex, and it has contributed significantly to Biogen's growth (it was delivered to market less than a year after FDA approval). For both the company and the young biotech industry, Avonex has become a symbol of breakthrough success. And people in

Biogen, one of the first biotech firms to inhabit the Kendall Square area, was the harbinger of a new industry, still closely linked to this neighborhood and the hospitals, universities, venture capital and research organizations located within a radius of several miles.

the industry believe this is just the beginning. According to a recent *Forbes* article, thanks to the growing biotech sector, a shift began in 2017 and 2018. While the average number of approved new molecular entities (NME) and new biologic license applications (BLA) for the pharma industry had been largely flat for many years, the industry rose significantly in those years, and the trend seems set to continue.

2004 In Remission
UNDERSTANDING HOW CANCER GROWS

Cancer remains perhaps the biggest challenge for medical research. Significant progress has been made in many areas of cancer research as doctors search for a way to defeat this multifaceted disease. One story of hope comes from the collaborative work done by two Massachusetts researchers, Dr. Judah Folkman and Robert Langer.

As a young and determined doctor, Folkman recognized something that others had overlooked: that all tumors depend on blood from the healthy portions of the body to survive and grow. He theorized that if researchers could find a way to discourage that growth, cancer could be resisted and perhaps defeated. At first, Folkman's theory was one element in a promising career that could have gone in more conventional directions. But with the help of Langer and his lab at MIT, Folkman's realization finally yielded a significant medical innovation.

Folkman first made observations that led to his theory when he performed research for the US Navy. He continued this research while practicing surgery and teaching pediatric surgery at Harvard Medical School. But when he suggested in the *New England Journal of Medicine* in 1971 that tumors actually attract blood vessels to grow toward them, a process he called angiogenesis, he was widely criticized or at best ignored. He further proposed that anti-angiogenesis substances might be discovered that could control or reverse cancer, but because his theories did not fit the existing ways of looking at the disease and because he was a surgeon, he was not recognized as a bona fide cancer researcher.

Undaunted, Folkman continued his work and began collaborating with Langer. Langer was an MIT chemist and engineer who focused not only on trying to identify substances with anti-angiogenesis properties but also on creating materials—typically polymers—that could be engineered to be semipermeable. In this way, the materials could provide a slow-release drug delivery mechanism within the body, which was potentially valuable for an anti-angiogenesis treatment.

Langer had little exposure to biology or medicine, but his skill and his fresh approach yielded results. Together, Folkman and Langer isolated from natural materials the first angiogenesis inhibitor, a macromolecule that could block the spread of blood vessels in tumors. This was the first real proof of Folkman's theory and a tantalizing hint that a treatment based on an anti-angiogenic substance might be a real possibility. Gradually, Folkman attracted more researchers to the field, and they eventually wrote thousands of reports in peer-reviewed journals. Others took notice. Monsanto signed an unprecedented agreement with Harvard to fund Folkman's cancer research and to seek patents to protect the work.

In 1981, Folkman resigned from his hospital work to focus on research. But despite methodical progress, convincing the medical

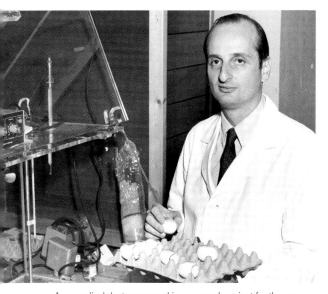

As a medical doctor engaged in a research project for the navy in the early 1960s, Judah Folkman observed that tumors couldn't grow without an adequate blood supply. This insight eventually led to his work on angiogenesis, which was seen for decades as impossible but helped produce Avastin, approved by the FDA as a treatment for colon cancer in 2004.

world that angiogenesis was potentially a key route to the control of cancer remained difficult. One of Folkman's more important experiments involved implanting a tumor in a rabbit's cornea, a tissue that normally has no blood vessels. Remarkably, after the tumor was implanted, blood vessels began to grow toward it. Folkman and his colleagues went on to test a wide range of anti-angiogenesis substances in mice, building up a deep body of knowledge and opening doors for other researchers.

In the meantime, Langer had achieved success with designing semipermeable polymers. His patented technology was used in the development of Norplant, for example, a long-term contraceptive that is delivered via implantable plastic capsules. From the late 1970s onward, Langer's MIT lab became (and remains) the largest biomedical engineering lab in the world, with scores of researchers funded by millions of dollars in grants.

Both Folkman and Langer continued to innovate, sometimes working collaboratively, and Folkman's dream became more achievable every day. Through the 1990s, new support began to emerge for his work—and results began to appear. One of his colleagues found that Thalidomide, a drug with an infamous association to birth defects, had important anti-angiogenesis properties.

Finally, in 2004, just a few years before his death from a heart attack, Folkman's long efforts were vindicated when researchers began clinical trials of an anti-angiogenesis drug (Avastin) that was soon shown to prolong successfully the lives of patients with terminal colon cancer. It was the first of a number of anti-angiogenesis drugs that have since come to market, all inspired by the pioneering work of Folkman and Langer. And while the angiogenesis inhibitors have not proven to be the silver bullet for stopping cancer, they have played an important role in the ongoing battle against the disease. They have also been applied to other conditions, such as macular degeneration, that involve the need to control the growth of blood vessels.

ENHANCING QUALITY OF LIFE

1820s

The Ice King
SELLING YANKEE ICE AROUND THE WORLD

On a hot day in July 1805, at an outdoor wedding party on the Maine coast, twenty-one-year-old entrepreneur Frederic Tudor and his brother William were cooling off with iced alcoholic drinks. Like all ice at the time, it had been cut by hand from the local pond during winter and dragged in blocks to an insulated icehouse for use in the warmer months. As they chatted, William remarked that people in the West Indies would give a lot to enjoy such a luxury. The guests laughed, knowing that ice would melt long before a ship could reach such a destination. But the comment got Frederic wondering how it might be done. He became so enamored of the idea that he traveled to Martinique four months later to begin marketing ice and iced drinks.

No one had ever shipped ice to the Caribbean, and it would be four years before Tudor would do so. He would turn a profit eventually by transporting Massachusetts ice in large wooden sailing ships, then selling it from warehouses he built. After years of experimentation, he worked out what type of insulation (typically some form of wood shavings) would be most effective at keeping ice frozen in the hold of a ship for a month or more and then learned how to transport the ice through the port to special double-walled warehouses in Havana, Cuba; Martinique; and New Orleans. Tudor worked with local taverns to determine what iced drinks would sell, and gradually he helped build up a culture of iced drinks at the ports of the Caribbean and the Deep South.

His first profit came in 1810, but two years later he lost a good deal of money to an unscrupulous agent. After spending part of 1813 in a Boston debtors' prison, he tried bringing oranges, lemons, and limes by ship to northern US ports. In spite of being encased in ice, however, the produce spoiled en route. Despite his many disasters, however, he would become one of America's first millionaires.

Tudor's crowning achievement was shipping ice to India—a four-month voyage. The ice helped develop a new tradition of iced drinks, including the gin and tonic invented by British soldiers stationed in India in the second quarter of the nineteenth century. Henry David Thoreau, in his cabin next to Walden Pond, loved the idea that his pond's ice was going into drinks for people in India, whose philosophy had so influenced him. Tudor's ice from a lake in Wenham, Mas-

The "Ice King," Frederic Tudor.

sachusetts, was very popular, by name ("Wenham ice"), in drinks at large parties in London society in the mid-nineteenth century.

Tudor employed whatever new technology he could to make the ice trade profitable. He built the first railroad in Cambridge and Boston to carry his ice from Cambridge's Fresh Pond through what is now Kendall Square to Tudor Wharf on Boston Harbor. He came to epitomize the innovative and imperious nouveau riche of Boston, New York, and the North in the 1830s and 1840s, and few could compete with the grandeur of what he called his "bold ideas." He worked closely with what we would call his operations director, Nathaniel J. Wyeth, to develop the best type of sawdust for maintaining cold temperatures in the holds of wooden ships.

Tudor was able to build local demand for his innovation by turning "crystal blocks of Yankee cold" into valuable products. He also operated on a global scale, using his entrepreneurial skills to build international demand for his innovation. Tudor Wharf still exists on Boston's HarborWalk, at the point where it enters Charlestown, and features images of the ice-cutting trade.

> Nathaniel Wyeth's son N.C., his grandson Andrew, and great-grandson Jamie used the ample family funds earned by Nathaniel's collaboration with Tudor to pursue quite a different path: all became great American painters.

During the nineteenth century, the cold winters of Massachusetts supported a vibrant ice industry.

1844

To Hell and Back

THE STORY OF CHARLES GOODYEAR

Few stories of entrepreneurship and innovation match that of Charles Goodyear's invention of vulcanized (all-weather) rubber. It is a tale that is as heart-wrenching as it is dramatic.

Goodyear was a thirty-three-year-old entrepreneur when he met with the owner of the Massachusetts-based Roxbury Rubber Company in the 1830s. Roxbury Rubber Company did a very good business manufacturing rubber goods with a recently developed British process that made India rubber for shoes and other clothing that could withstand the cool, rainy British weather. Goodyear had a proposal for improving life preservers, one of the earliest rubberized goods, but he had approached the company at the wrong time. In the hot, steamy Boston summers, rubber shoes that could withstand the cool weather in Britain turned to a messy goo, and in the cold Boston winters, the rubber turned hard and crumbled. Customers from across the region had been returning their ruined purchases, and Roxbury Rubber, like many US companies making rubber products, was about to go bankrupt.

Goodyear saw an opportunity. He told the owner of Roxbury Rubber that he would work on this problem. But he had business problems of his own, in the form of a store in Philadelphia that was itself slipping into bankruptcy. Goodyear had an earlier career making improvements in implements, so while in debtors' prison for that bankruptcy, he turned his inventive mind to the challenge of India rubber, warming it on a flame and adding magnesium oxide. He felt close to a solution, and friends raised money to get him out of prison.

He went to New York City to continue his experiments. He sold his furniture, sent his family to a quiet boarding place, and set himself up in an attic. There, with help from a druggist who was knowledgeable in chemistry, he continued his experiments. He compounded rubber with magnesium oxide and then boiled it in alkali and water. This process seemed to have made the rubber much less sticky, and he was applauded internationally for his success.

But then Goodyear saw that a drop of weak acid falling on the cloth neutralized the alkali, immediately causing the rubber to become soft again. He found that magnesium didn't solve the problem, but he remained obsessed with finding a process that would create pliable, waterproof, moldable rubber—"vulcanization." But Goodyear's personal troubles

> "Life should not be estimated exclusively by the standard of dollars and cents. I am not disposed to complain that I have planted and others have gathered the fruits. A man has cause for regret only when he sows and no one reaps."
>
> —Charles Goodyear

Charles Goodyear's "eureka moment" came after years of dogged perseverance that brought him closer and closer to his breakthrough.

continued. Harsh chemicals, including nitric acid and lead oxide, ruined his health and once led to a fever that nearly killed him. After the financial Panic of 1837, he left New York and moved with his family to Woburn, Massachusetts, a farm town outside Boston. For another five years, he worked on the process while his family grew destitute. He was so poor that he often couldn't buy food for his wife and children. He returned to debtors' prisons for years at a time but worked more on the rubber invention in his cell. His brother-in-law, seeing Goodyear's situation, told him that "rubber is dead."

"I am the man to bring it back to life," replied Goodyear. An extreme of the model of the innovative entrepreneur, Goodyear had the desire, so pervasive in Boston's DNA, to create this world-changing innovation. He was willing to go to debtors' prison himself for years, as long as he could learn how to vulcanize rubber.

The big discovery finally came in the winter of 1839. Goodyear had been experimenting with sulfur, and the story is that one February day, he wandered into Woburn's general store to show off his latest gum-and-sulfur formula. Snickers rose from the local farmers sitting on the store's barrels; they had heard many of Goodyear's claims of progress. Goaded on, perhaps, the usually mild-mannered inventor got excited and waved his sticky fistful of gum in the air. It flew from his fingers and landed on the sizzling-hot potbellied stove in the store. When he bent to scrape it off, he found that, instead of melting like molasses, it had charred a bit but was tough and flexible like leather. And around the charred area was a dry, springy brown rim; it was gum elastic still but so remarkably altered that it was a new substance: weatherproof rubber.

This discovery is often cited as one of history's most celebrated accidents, but Goodyear stoutly denied that. Like Newton's falling apple, he maintained, the hot stove incident held meaning only for the man "whose mind was prepared to draw an inference." He maintained that he was the one who had "applied himself most perseveringly to the subject." Goodyear did not act alone, however; he had been in constant contact with others working on the chemistry of substances that might be able to vulcanize rubber.

The story was not over. Goodyear started a small business to make rubber shoes. They were better than earlier products, but the quality was still not good enough and many were still returned. For four more years, he persisted in Woburn, and then he started a small factory in Springfield, Massachusetts. Finally, in 1844, he patented his vulcanization process, and his Springfield factory started producing goods that worked.

Today Goodyear is a household word. He *had* brought rubber back to life! But after all that heartache, was it instant success? No, seven years of patent fights in the United States and in England followed. Although he had a good case that an English company had patented a formula copied from his, Goodyear lost the case in English courts. In the United States, where the English firm sought a patent, Goodyear won. Although his Springfield factory was successful, and although an Ohio manufacturer of rubber tires would later immortalize the Goodyear name, the cost of marketing combined with his legal fights overwhelmed him, and he died penniless in 1860.

1846

A Stitch in (Less) Time

THE SEWING MACHINE REVOLUTION

Mechanical aids for spinning thread and yarn (the spinning wheel), and weaving cloth (the loom) date back to antiquity. Illustrations of looms have been found among the artifacts of ancient Egypt, and archaeologists believe the technology was old even then. One of the early fruits of the industrial revolution was adding power to these traditional tools and refining them further so that thread and cloth soon became commodities.

What remained unchanged as of the middle of the nineteenth century was the actual making of clothes, still a painstaking process built around innumerable hand stitches, work that was usually relegated to women. A number of inventors had tried to improve this process by mechanical means but with little success. Elias Howe, a country boy who had learned about textiles as an apprentice in America's first textile city, Lowell, Massachusetts, and later as a mechanic for a company that made carding machines in Cambridge, Massachusetts, changed that pattern of failure.

Earlier attempts to mechanize sewing imitated the dance of needle and fingers that a human performs to join pieces of cloth. In contrast, Howe absorbed concepts from the fertile industrial environment of the time, including the manufacture of scientific instruments and loom building. Howe placed the loop of the needle near the pointed tip and included a shuttle and an additional spool of thread below the cloth. The machine inserted the needle and thread into the cloth, and a shuttle, located below the cloth, seized this loop of thread and pulled a second thread through it, creating a lockstitch. The primary characteristic of a lockstitch is the linking of the two threads, accomplishing results that are similar to what a human can do with single thread and needle but at a blindingly fast pace. The third element of the design was a mechanical method to advance the cloth automatically with the completion of each stitch, positioning the cloth for the next stroke of the needle.

Elias Howe learned much of machines and business from the community that surrounded him. He grasped the potential value of eliminating the drudgery of thousands of hand stitches required to make a garment and created a true breakthrough.

This singular invention, perfected at Howe's home in Central Square, Cambridge (on Cherry Street, the same street as the home of transcendentalist Margaret Fuller), was awarded a patent in 1846. But Howe's machine was not an immediate success; he initially lacked sufficient capital to manufacture it and had difficulty attracting any US investors. His elder brother, Amasa Bemis Howe, even traveled to London in search of backing. In the meantime, his prospects were marred by rivals such as Isaac Singer, whose machines copied Howe's lockstitch idea. Although Howe eventually won a lawsuit against Singer, forcing him to pay royalties, the legal challenge kept him from manufacturing on a large scale until 1863, when he formed a company and built a large sewing machine factory at Bridgeport, Connecticut.

Although Howe's designs eventually received many international awards, Singer proved more adept at selling and marketing his machines, lowering prices to make them more and more affordable and adding features. Singer also manufactured them in multiple locations in the United States and abroad, eventually dominating the market. Today, the Singer business continues to be an important global brand.

Howe's firm prospered for a time. He also patented a pioneering design for a "continuous clothing closure"—what we would today call a zipper—in 1851, but he made little effort to develop the idea, and it was largely forgotten. Howe died in 1867 at age forty-eight. He had succeeded in inventing, making, and selling the first sewing machines. At first, he had to overcome unfamiliarity with the new invention among the public and initially, as with many new innovations, the cost was high, limiting sales. As has happened with so many other innovators over the years, the "first" is often overtaken by later companies with better funding and more marketing know-how. Singer and others reaped more profit, but it was Howe and his sewing machine—even memorialized on a US postage stamp—that paved the way.

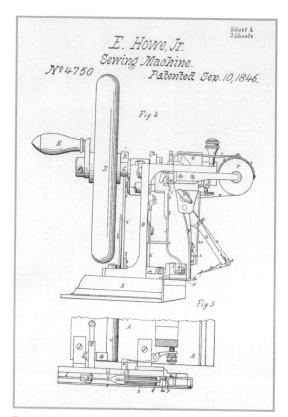

The drawing of the sewing machine that Howe prepared for submittal to the US Patent Office.

1904 Cutting Edge
KING GILLETTE AND THE DISPOSABLE SAFETY RAZOR

For many, the name Gillette is synonymous with shaving. It was the surname of a great salesman and inventor, a man with a big idea who lived in Boston's Fenway neighborhood. Gillette not only sought a fortune but also originally wanted the profits of his vast business success to fund a utopian socialist city and create peace in the world. Perhaps it should be no surprise that his first name was King.

In 1895, when Gillette conceptualized his razor, men shaved with straight razors. Frequent, time-consuming honing and stropping were required to keep the blade of a straight razor as sharp as possible. Most men in the United States shaved every other day, and many grew beards. Gillette's big idea was to use the features of the newly developed safety razor, which made it almost impossible to make a deep cut or gash, and equip it with a blade that was so inexpensive it could be thrown away after a couple of days' use. Gillette's concept had two potential advantages: it eliminated blade sharpening, often done at a neighborhood barbershop, and so made shaving much more convenient; and it meant that an established customer would have to keep buying Gillette blades once he had purchased the inexpensive Gillette razor handle. Creating an innovation takes time, however, and it would be eight years before Gillette's new idea hit the market.

Before coming to Boston from his native Wisconsin, Gillette had worked as a salesman for Crown Cork & Seal, which had developed disposable caps for bottles. Gillette wanted to extend the idea of disposability to shaving, but one of his biggest problems was constructing a working model. He had the vision but not the technical knowledge to craft a suitable blade. He needed to find an engineer who could work out how to create

King Gillette revolutionized the way men shaved. He fostered both breakthroughs in manufacturing and mass production as well as approaches to marketing. The company he founded is now part of Procter & Gamble and is one of the most recognized global brands.

a wafer-thin blade from sheet steel and then harden it so that it could hold a sharp edge.

Experts at MIT told him that was impossible; only forged steel, they said, could hold the sharp edges necessary for shaving. When he pressed them, they told him to speak to another MIT engineer named William Nickerson, who had some unique ideas but was something of an outcast. It was his pursuit of this unpopular engineer that led to Gillette's success. After examining Gillette's mockup for thirty days, Nickerson figured out a way to manufacture a blade with the right consistency and told Gillette that it could be done. Nickerson experimented with various types of abrasive wheels, cutting angles, and grinding speeds. By 1902, seven years after Gillette's first conception, he finally developed a working model of a disposable, double-edged razor blade that didn't need to be sent out for sharpening.

As so often happens with a start-up, however, Gillette did not have the funds to start production on a large scale, though he did start producing some razors in 1903 at a little place above the fish market next to Boston's South Station. He and several business friends approached John Joyce, who had invested funds in a previous Gillette invention that failed. Reluctantly, Joyce agreed to invest but required that Gillette give up a substantial part of his new company to the financier. By the end of 1903, fifty-one razors and 168 blades had been sold; the next year, sales increased to 90,000 razors and 124,000 blades. Gillette, the company and the man, were on their way to fame and some fortune.

To counter the fear that his razor was an "open sharp knife," Gillette used this ad of a baby shaving to emphasize his invention's safety.

Gillette and Joyce focused the company on two marketing concepts: changing the culture of shaving to a daily activity and expanding globally. The latter was still unusual for US consumer product companies in the early 1900s. Their advertisements emphasized the manliness and attractiveness of a smooth-shaven man, a message they enhanced with a new form of advertising: signing Major League Baseball players such as Honus Wagner to endorse the razors. "Your baseball star . . . starts the day with a clean shave—and, like all self-reliant men, he shaves himself," the advertisements claimed.

But it took World War I to secure mass-market success for home shaving. In 1918, the US military started to issue Gillette shaving kits to every serviceman. Gillette sales quadrupled. The company began to produce blades in Paris and other cities around the world, and within two decades, Gillette's image, which graced every packet of the company's razor blades, was perhaps the best known in the world.

1930
Inventing America's Favorite Cookie

A world without chocolate chip cookies? Unthinkable. The chocolate chip cookie is the most popular cookie in the United States. But that wasn't always true. And its invention could well have been a mistake. Or was it an experiment that helped a keen-minded nutritionist learn from an unintended outcome?

After studying nutrition at Framingham State Normal School (now Framingham State University), Ruth Wakefield and her husband bought an old house used to collect tolls along a former turnpike route in Whitman, just south of Boston, and turned it into the Toll House Inn. She was the baker and chef. Her incredible desserts began attracting people from all over Greater Boston, and one day in 1930, while experimenting with the butter drop do, a favorite colonial-era cookie, Wakefield used an ice pick to break a bar of ice-cold semisweet chocolate into pea-sized bits and added them to her cookie dough. She had planned to melt the chocolate in advance, but she was rushed and hoped the bits would melt during baking so that a chocolate cookie would emerge.

However, she was using Nestlé chocolate—the only chocolate she happened to have on hand that day. Nestlé chocolate had a different formulation from the Baker's Chocolate she usually used. Instead of melting, the chocolate pieces held their shape after the cookies had baked. She served them that day for those wanting a new type of chocolate cookie, and

The original Toll House, where the chocolate chip cookie was invented.

Innovations which change the world are not limited to technology and medicine. The now ubiquitous chocolate chip cookie started at a small restaurant near Boston in the 1930s. Known as the Toll House, its Toll House cookie recipe launched the chocolate chip craze.

the customers loved them. She called them the Toll House chocolate crunch cookie. As the cookies became the most-requested item on the menu, she shortened the name to Toll House cookie.

During the 1930s, the cookies became a local phenomenon. Nestlé made a deal with Wakefield to supply her with a lifetime of free chocolate in return for the right to publicize the recipe and to use the Toll House name. Later Nestlé began to score its slabs of semi-sweet baking chocolate to help those trying to duplicate Wakefield's recipe. Before long, the company took the next step of marketing the chips as "chocolate morsels," which soon changed to "chocolate chips." During World War II, as soldiers from Massachusetts often shared cookies sent from home, the popularity of the recipe grew beyond its local region.

Was this innovation just luck? Not really. Ruth Wakefield was a trained nutritionist and a perfectionist. She was looking for scrumptious new items for dessert to vary her menu. She took advantage of a new supplier and adapted quickly to positive feedback from her customers.

> The most popular cookie in the United States back in 1930—before the advent of the Toll House cookie—was oatmeal raisin.

1930s
Fresh Frozen
THE BIRDSEYE WAY

Going fishing is not an activity most people think of as entrepreneurial. Sometimes it is even a synonym for laziness. But for Clarence Birdseye II, fishing was the key to a world-changing innovation and personal wealth.

Birdseye attended Amherst College in Massachusetts before moving west to work for the US Department of Agriculture in support of the researcher who eventually discovered the cause of Rocky Mountain spotted fever. He then pursued a series of research projects in Labrador, where local Inuit people taught him the art of fishing through extremely thick winter ice, with atmospheric temperatures often as low as $-40°F$. To his amazement, he saw that fish caught in that frigid environment froze solid almost immediately. Even more striking was that when they were later thawed and cooked, the texture and flavor were almost indistinguishable from fresh-caught fish.

These observations convinced him that he might have found the key to improving frozen food processing: faster freezing and even lower temperatures. Of course, many people, especially in colder climates, were already familiar with frozen food, either frozen inadvertently and packed in ice or artificially frozen commercially. Unlike the fish Birdseye caught and froze in Labrador, however, the results upon thawing tended to be disappointing—either too mushy or too dry.

After some promising experiments, Birdseye established a business, General Seafood Corporation, to focus on fish, a famously perishable product. As home base, he chose one of the nation's leading fishing ports, Gloucester, Massachusetts, where he used his latest invention, the double-belt freezer, to freeze the product rapidly between a pair of moving steel belts chilled to a very low temperature.

The company soon had a faithful customer base as Birdseye's inventive mind added a series of improvements to the process. He began to freeze other meat and vegetable products, and General Seafood became so successful that Birdseye was able to sell the company for the fantastic sum of $22 million in 1929 (the equivalent of over $300 million today). The purchasers eventually formed the Birds Eye Frozen Foods Company and Birdseye himself stayed on to continue to improve the technology.

In 1930, Birdseye implemented the first large-scale experiment to sell frozen foods directly to the public. He worked with eighteen retailers in the vicinity of Springfield, Massachusetts, and helped develop the display cases similar in concept to those still in use today. Despite it being one of the worst years of the Great Depression, consumers proved more than willing to try the new products and came back for more, laying the foundation for today's global frozen food market. A taste for frozen foods has since become a worldwide phenomenon. In the United States alone, according to a recent report by Grand

Frozen foods, first widely test marketed in Springfield, Massachusetts, have become a global industry.

View Research, the frozen food market is expected to reach a value of nearly $73 billion by 2024.

After making his breakthrough, Birdseye's research revealed that fast freezing works because it produces much smaller ice crystals that are less likely to rupture individual cells or damage tissue. Today's industry remains on the path he established, seeking ways to speed the freezing process even more. Birdseye, who died in 1956, asked that his ashes be scattered in the sea off the coast of Gloucester. In 2005, he was inducted posthumously into the National Inventors Hall of Fame.

1946 Hot Stuff

FROM THE RADARANGE TO THE MICROWAVE

In 1945, a Massachusetts inventor, Percy Spencer, at the Raytheon Company was fresh off a career helping the Allies win World War II by mass-producing the revolutionary magnetron, the key component in radar technology. Spencer changed cooking history when he turned his attention to a candy bar. After lunch, Spencer often bought a chocolate bar to give him energy later in the afternoon. One day he discovered that the chocolate in his jacket pocket had melted after he'd been standing next to a magnetron, which emanated concentrated energy in the form of invisible, high-frequency radio waves. If those waves could melt his candy bar in seconds, he reasoned, they should quickly heat and cook other foods. He bought some unpopped popping corn and put the bag in front of the magnetron, and he had popcorn. It worked.

British scientists had invented the magnetron and radar. During the Battle of Britain, with many factories bombed, the UK government needed to change the laborious process of handcrafting the copper magnetron. They approached Cambridge-based Raytheon and met Percy Spencer. He came up with a solution to the radar manufacturing challenge over a weekend at his kitchen table, and Raytheon was hired to mass-produce the new radar equipment in Newton and Waltham, Massachusetts, at factories far outside the range of Luftwaffe bombs. But by the end of the war, Raytheon and its top inventor were thinking about how to switch from war to peacetime production.

Spencer and his Raytheon colleagues worked on developing the Radarange, filed a patent application, and within a year produced the first microwaves built around the magnetron. By 1947, they offered the first commercial microwave oven, dubbed the Radarange. But the innovation was not yet a home appliance. Adapting the same complex and powerful circuits used to detect approaching enemy aircraft resulted in an appliance the size of a refrigerator, weighing several hundred pounds and costing thousands of dollars.

> The son of a Maine millworker, Spencer was a self-taught engineer with little formal schooling. He completed only the third grade. After serving in World War I, he was homeless before being hired by Raytheon. He would earn a total of nearly three hundred patents over the course of his career, but his perfection of the magnetron and its application to cooking was his most lasting legacy.

A chef posing with a 1940s Radarange, the first commercial microwave oven.

cooking. A microwave oven was put into service at Norumbega Park, for example, a suburban Boston amusement park, where it wowed the crowds by serving up freshly made popcorn in seconds.

Design refinements, production improvements, and Raytheon's acquisition of the appliance maker Amana helped bring the price of a Radarange under five hundred dollars by the late 1960s, which was still a lot of money to most home users. But over the coming decade, further product improvements and competition cut cost and size and added features. US microwave oven sales rose from forty thousand units in 1970 to one million five years later. Global sales also rose, making the microwave oven commonplace in most developed world homes by the 1990s.

Given the cost and size of the product, its initial applications were for a few commercial establishments that had time-sensitive customers or an interest in the novelty of instant

Today the microwave oven is used wherever electricity is available. And its energy-efficient cooking also means that the world can prepare food for more people using fewer natural resources.

1960 The Pill
PUTTING WOMEN IN CONTROL OF THEIR FERTILITY

Few innovations have affected the lives of ordinary Americans as much as the birth control pill. And few have been as controversial. By the 1940s, inexpensive synthetic hormones had made it clear to many in the scientific community that it was possible to manipulate human fertility. Despite the great potential market for doing so, neither pharmaceutical companies nor researchers seemed prepared to take that step. Part of the reluctance stemmed from conservative attitudes among some sections of the population as well as numerous statutes, often ignored in practice, that made any kind of contraception illegal.

In the early 1950s, that lack of change was ended by a handful of individuals linked through institutions, traditions, and people within Massachusetts. The people involved included Katharine Dexter McCormick, an heiress with a degree in biology from MIT; Margaret Sanger, a famous campaigner for family planning; Dr. Gregory Pincus, Dr. Min Chueh Chang, and Dr. John Rock, fertility researchers in Boston and Worcester; and, much later, Bill Baird, a civic activist determined to realign the legal framework surrounding reproductive rights.

Dr. Pincus, cofounder of the recently established Worcester Foundation for Experimental Biology (WFEB), met outspoken birth-control advocate Margaret Sanger, who had been promoting efforts to give women more control over their reproductive lives for more than thirty-five years, often in defiance of the law. Also at that meeting was Abraham Stone of Planned Parenthood, who had begun to provide funding to help Pincus with his hormonal research, which had potential for contraceptive application. Pincus had been working closely at WFEB with Dr. Min Chueh Chang, a biologist who had been educated in his native country of China and in the United Kingdom, and was a recognized expert in mammalian reproduction. Although G.D. Searle, a leading drug company, agreed to supply the project with chemicals, they

Social activist and suffragist Katharine McCormick played a crucial role in providing financial support for research that led to development of the Pill.

Margaret Sanger, whose name was synonymous with birth control in the US throughout much of the twentieth century, saw the potential of research underway in Massachusetts and helped engage financial support from Katharine McCormick.

Progress was rapid, and clinical trials began in 1956 on a combined oral contraceptive that included noretynodrel and mestranol. Manufactured by G.D. Searle, the product was called Enovid. But it wasn't *the Pill* yet. Approved by the FDA in 1957, it was reserved for use in treating menstrual disorders. In fact, it wasn't until 1960 that the FDA granted approval for it to be prescribed as a contraceptive, and in many states, it remained illegal or available only to married women.

The Pill, as it became almost universally known, rose in popularity: more than a million American women had prescriptions by 1962 and another million joined them within a year. It remained controversial and hard to

Dr. Min Chueh Chang, a member of the Worcester Foundation team and a leading reproductive biologist, provided insights that were crucial to development of the oral contraceptive.

declined at that point to provide financial support.

Impressed by the progress made by Pincus and Chang, Sanger introduced Pincus to her friend Katharine Dexter McCormick, a philanthropist, women's rights activist, and MIT graduate. McCormick decided to increase funding for the work at WFEB dramatically. Pincus also reached out to Dr. John Rock, chief of gynecology at the pioneering Free Hospital for Women (which operated in Boston from 1875 to 1965) and a Harvard medical school professor, to undertake clinical studies.

In 1964, Dr. Gregory Pincus (right), was honored by Dr. Paul Wermer, medical director of the City of Hope Medical Center, "in recognition for his fundamental research on hormones leading to the products now used in prevention of conception."

obtain, however, even in the state where it was developed. Nor was the controversy confined to the United States. After convening a special commission to consider the moral merits of the Pill, the Catholic Church pondered the matter and ultimately declared its opposition. Meanwhile, a Connecticut court case, *Griswold v. Connecticut*, which eventually reached the Supreme Court in 1965, was decided in favor of the plaintiff, effectively making the Pill legal for *married* women in all fifty states.

The invention and commercialization of the oral contraceptive "the Pill" helped launch a social and sexual revolution.

In Massachusetts, a vociferous activist, Bill Baird, had been petitioned by students at Boston University to challenge the state's archaic Crimes against Chastity, Decency, Morality and Good Order law. In 1967, he gave a speech about birth control on campus and gave contraceptive materials to a woman—and was promptly arrested. Facing up to ten years in jail, he was sentenced to only three months, but through appeals, his case also came before the US Supreme Court. In 1972, the court decided in his favor on the basis of its expanding interpretations of a right to privacy, a decision that effectively ended any prohibitions on adult use of the Pill or any other contraceptives outside marriage. In most parts of the world, the Pill and many other forms of contraceptive are now widely available.

DOING BUSINESS TAKES MONEY

1784 Building America's Lending

THE NATION'S FIRST CHARTERED COMMERCIAL BANK

One year after the signing of the peace treaty that ended the American Revolution, several Massachusetts patriot leaders and merchants, led by the merchant and governor John Hancock, sought to establish a bank to help rebuild an economy decimated by the war. The British occupation of Boston, the nation's third largest town, had resulted in the loss of 85 percent of its population. Prior to the Revolution, Boston had a strong economy due in part to its large trade with the British West Indies, but for a decade the town had been cut off from meaningful trade.

In 1784, the Massachusetts state legislature granted a petition from Hancock and five other Boston merchants to incorporate a commercial bank with what was then the huge sum of over $250,000. Incorporated by an act of the legislature, the General Court of Massachusetts, the Massachusetts Bank was the first bank in the new nation to have a state charter, implying a degree of public oversight and thereby giving its loans credibility. The bank raised capital by selling shares and earned revenue by charging a small sum for deposits and discounting notes. It is the earliest direct ancestor of today's mammoth Bank of America.

One of the bank's early projects was to find an outlet for US goods to help Boston's economic recovery. The bank played a large role in putting together funds for the 1787–1790 trip by the ship *Columbia* to the virgin Pacific Northwest, where the crew purchased otter pelts from Native Americans and then brought them to China, where the pelts were traded for silk, tea, and other goods. The trip was an important part of building new trade routes, and Boston and its local rival Salem would go

This medallion was struck to commemorate the *Columbia*'s achievement, as well as that of the *Lady Washington*, a smaller vessel that accompanied *Columbia*.

The *Columbia*'s voyage to the Pacific Northwest, financed in part by the nation's first chartered bank, launched Boston's pelt trade with China and other parts of the globe.

on to develop lucrative trade with China in the 1790s.

Developing and chartering a bank was a major innovation. For 130 years up to the American Revolution, the British had effectively stopped a number of Massachusetts business leaders who had been focused on starting American homegrown banks or financial institutions to help merchants finance large business deals. In 1686, there had been a public land bank in Massachusetts. Although little is known about it, this bank appears to have issued paper money based on land and goods. Plans were drawn in 1713 for a second land bank, but the colonial government forbade it, and by 1741, all banking was forbidden by the colonial government and the British Parliament.

Before 1789 and the development of the US Constitution, the national government had no way to be involved in regulating banking, which would give institutions an ability to loan and hold funds at a higher level of credibility. During the 1780s, only an independent state could step in and fill this function. Seeing the need, the state of Massachusetts explored the idea of having a bank with some government oversight and thus granted a state charter to a commercial bank.

The importance of this act should not be underestimated. As the economic historian Howard Bodenhorn has put it, a charter was "nothing less than a legislative imprimatur. It served notice that someone had assessed the capabilities of the applicants and had found them adequate." By 1790, there were three chartered banks in the United States: in Philadelphia, Boston, and Baltimore; by 1800, there were twenty-eight state-chartered banks, with the number growing exponentially and state banking regulation becoming the norm for many states.

1946 | Fueling the Start-Up Economy

THE FIRST MODERN VENTURE CAPITAL FUND

Venture capital is the lifeblood of innovation. It is vital for start-ups, for growth, for most new innovative businesses, and for new sectors of the economy. The incredible wave of innovation that swept through Massachusetts in the late twentieth century—from electronics and computing in the 1980s to biotech in the 1990s and beyond—produced hundreds of thousands of new jobs in thousands of companies. That growth would not have happened without venture capital.

The riskiest period for a business financially is the start-up phase. Although a new business may have a brilliant new product or service, a viable market may not exist for the new product or service. New companies lack a track record of sales and performance. For high-growth technology industries such as biotech and app creation, the risk is especially evident. In many cases, these new businesses are creating value with something so new that it is impossible to quantify returns with any degree of certainty.

Most banks and many investors are reluctant to fund start-up companies under such circumstances. The small number of wealthy investors who understand the risks and take the plunge with start-ups are known as angel investors. Beyond the angel investors, however, venture capital (VC) firms are set up specifically to target such new companies with special funds developed for these business opportunities; they take over from the angel investors. These VC firms pool funds from investors and institutions that are a bit less tolerant of risk than the angel investors but are nevertheless willing to gamble in order to reap great rewards. If and when the new company has grown to a point where it has enough regular sales of its products or services to make a profit, it sells shares and repays the VC firm for the money it invested—plus a large payoff for taking the risk. (It is important to note that many start-ups do not succeed, and venture capital firms can also lose a good deal of their money.)

Before the advent of modern venture capital, innovative ideas were regularly turned down by investment houses. It is true that venture capital has always existed in some form, going back at least to the Medicis in Florence in the fourteenth century, when someone with a new idea would approach a wealthy family for funds to pursue the new idea until it could become profitable. But it was not a formal process or system, and many potential innovators did not have wealthy connections.

Modern venture capital dates to the end of World War II, when US economic leaders were concerned that the nation would return to economic depression without something to add dynamism to the economy. Under the impetus of war, researchers had made giant strides in many fields, and these leaders sought a way of harnessing that creativity and turning it into new peacetime businesses. The Harvard Business School professor Georges Doriot was one who sensed that, without investors looking to fund new, risky inventions and start-up companies, great opportunities would be lost, and job creation would stay stagnant, perhaps pushing the US economy back into the kind of depression it had suffered before the war.

In 1946, Doriot founded America Research and Development (ARD), the first formal venture capital company, and raised $5 million to invest in new ideas. Doriot's idea was that a new venture capital firm would raise funds from wealthy individuals who would pool their resources and, after listening to entrepreneurs pitch their business ideas, might choose to invest in new companies. The venture capital firm would often take a seat on the board of directors of the company it had invested in to provide business expertise and increase the chances of the new company's success. The board of directors Doriot built for ARD was filled from Boston's financial and academic elite, including the presidents of MIT and the Federal Reserve Bank of Boston. Doriot ran the firm out of his house next to Boston's Beacon Hill.

ARD's first investment was in High Voltage Engineering Company, a firm that used X-rays to treat cancer. After eight years, investors were rewarded with ten dollars for every dollar they had invested. Another early investment, Digital Equipment Corporation (started by MIT computer specialists), saw ARD's investment of $70,000 grow to $183 million by 1967. By the 1980s, Digital Equipment Corporation employed 120,000 in the Greater Boston area (before it missed the personal computer wave and disappeared).

Like the innovators behind the telephone and clipper ship industries, Doriot was an immigrant who had come to the United States to do something for his future. Born in France in 1899, Doriot was schooled in innovation by his father, who helped design the first Peugeot automobile. After World War I, his father funded Georges's study in London and then at Harvard Business School, where he earned an MBA. His subsequent brief career as an investment banker is said to have turned him off the conventional methods of commercial

American Research and Development aimed to help spawn new companies and industries. By 1953, the cover of its annual report, consisting of a large collection of logos representing its diverse investments, testified to the success of its efforts.

investment. He felt that there was a big disconnect between the innovative teachings of "idea investment" at Harvard and its application in the financial world of 1920s Boston and New York.

Doriot started thinking about the concept of venture capital. Harvard hired him to teach courses about manufacturing and industrial management at Harvard Business School, and he also took on CEO and chair of the board responsibilities for several companies. After becoming a US citizen in 1940, he resigned his corporate positions and was made a lieutenant colonel in the Army Quartermaster Corps. Here, he initiated a number of innovative organizational changes and pursued research-based improvements in gear, notably boots that were appropriate for the many different environments in which US soldiers were deployed, ranging from steamy jungles to arctic cold.

After spearheading the first formal venture capital firm, Doriot ushered the new industry past a number of large obstacles during its first three decades. Many financial regulations and tax laws made it difficult to operate in the freewheeling way that venture capital needed in order to transform new ideas into new companies and to survive long enough to become important contributing firms in new industries. Among other barriers, the US Securities and Exchange Commission (SEC) regulations limited any ARD board member from getting stock options. SEC investigators also argued that the time frame—a minimum of ten years for the investments to provide a return—was too long. Doriot argued that a new firm with a new idea easily takes a decade to become stable and profitable.

In 1972, after twenty-five years, Doriot closed ARD as a separate company and folded it into a large company, Textron. It took until the late 1970s, during the Carter administration, to ease regulatory strictures and expand the playing field for venture capital firms. As a result, the industry blossomed, particularly in California. Today, measured by the value of funds invested, California is the largest recipient of venture capital funds, with New York second and Massachusetts third.

One of the five top drivers of Boston's success in innovation has been the availability of local funding to launch new businesses and to provide additional capital to fund later rounds of growth. Doriot played a leading role in helping the new industry of venture capital get started, and venture capital is a crucial factor in the creation and maintenance of an innovation-oriented economy in the Greater Boston area.

1974
Mutually Beneficial
GIVING MUTUAL FUND CUSTOMERS QUICK ACCESS TO THEIR CASH

Innovation is often the result of suggesting something that is counterintuitive. Such suggestions can reveal new ways of doing something different that people hadn't thought of because they were doing what they had always done in the way they had always done it. And yet, when the change is suggested, it seems so obvious.

Fidelity Investments CEO Ned Johnson suggested a counterintuitive change when he imagined that if mutual fund shareowners could write checks to take money out of their accounts, they would be more willing to deposit money. This simple innovation led to increased market share for Fidelity Investments, producing $7.5 billion in new funds and, along with other financial innovations, made Fidelity Investments the largest retail money manager in the world in the 1980s and 1990s.

Johnson's innovation was in a long tradition of financial inventiveness in Massachusetts. In the 1820s, Massachusetts financiers delineated the key tenets of how to save family funds by setting up trusts. With trusts, it was possible for the families who had made money in New England's early industries to live off their savings for four or five generations. Greater Boston became known nationally for its investment expertise and leadership.

A century later, Boston financiers invented the mutual fund, another mechanism for those with money to invest in the stock market by pooling funds to limit risk. Three different Boston firms developed the mutual fund in the 1920s. Run conservatively, in a city dominated by old money, the mutual fund was intended primarily to avoid losses and achieve reasonable gains. It was a very conservative financial instrument.

Fidelity Investments Inc. widely advertised their "Fidelity Daily Income Fund," a mutual fund that allowed middle-class consumers to deposit their excess funds and get daily interest.

Johnson was a Boston conservative in tone and manner, but he was not conservative about money management. The point of mutual funds for him was not just to preserve capital but to add to it. By the 1970s, when Johnson was pondering how Fidelity

Investments might reform mutual fund strategy, the mutual fund marketplace was getting very crowded, very quickly. A large Boston fund, the Dreyfus Corporation, was waging expensive advertising campaigns that Fidelity Investments couldn't match. Johnson decided to fight back.

Surrounding himself with highly capable individuals and learning from them, Johnson used check writing on mutual funds to increase fund flow dramatically, and Fidelity became the largest mutual fund owner in the world. Earlier, stockbroker Bernie Cornfield had introduced check writing on relatively small funds, but Fidelity Investments was the first to grant check-writing privileges to its customers on a large scale. Johnson's team included experts in different but equally necessary aspects of the mutual funds business. He was able to draw out from his team members a major change in strategy that increased Fidelity Investments' assets by $7.5 billion.

With the increase of assets by other Boston-based investment firms, including Putnam, Massachusetts Financial Services (MFS), and Pioneer, Boston financial institutions had more assets under management by the end of the twentieth century than any other city in the world. Boston remains one of the top globally.

Ned Johnson's innovation of allowing the writing of checks against mutual funds attracted a great deal of money to Boston in the 1980s and 90s, making it the number-one city in the world for asset management of mutual funds.

NINETEENTH-CENTURY HIGH TECH

1836 Digging In
THE FIRST POWER EXCAVATOR

The first half of the nineteenth century was a period of major innovation in global transportation. In particular, the railroad transformed the way people and

> William Otis should not be confused with his distant younger cousin Elisha Otis, who developed a critical part of the elevator and started the Otis Elevator Company.

goods were moved, which in turn had a major impact on the world's industries, including the development of construction equipment. But the transformation worked both ways: the invention of the steam shovel by William Otis, a twenty-year-old man from Canton, Massachusetts, would influence how the railroads themselves were built and later all kinds of construction.

William Otis had been working for a railroad construction company that promised bonuses for additional excavated loads. Otis and his team were tired of the slow pace of their tools. The old-style wagon-mounted graders and horse-drawn dragpans—and manual labor—inspired the young man to invent the steam-powered shovel, the first successful power excavator. The needs of railroad builders directly influenced Otis's invention. The disruptive wave of steam-powered machines that led to the first railroads in the United States in the decade prior to Otis's invention led him to persist in overcoming numerous obstacles in developing his new machine in the small industrial town of Canton, south of Boston.

The industrial environment of Canton, and the random conversations that were possible in such an environment, provided crucial support for Otis as he perfected his invention. In particular, conversations with the blacksmiths in their sheds around the Revere Copper Company (started originally by Paul Revere) made the difference in connecting the heavy metal shovel with the arms of the new steam machine. Among other things, the plant manufactured the large sheets of copper used to protect wooden-hulled ships like the *USS Constitution*. Blacksmiths in adjoining facilities worked with iron. Otis depended on the ideas and comments they provided as he experimented with the early design of his steam shovel. The combination of a driven entrepreneur with an informal network of people different from his experience and mindset, in this case, blacksmiths and coppersmiths, resulted in a critical phase in the invention: the connection of the big bucket to the other parts of the excavator.

In Otis's first machine, the hoisting of the bucket was the only action powered by steam. The machine itself was moved by a team of horses, and men used ropes to swing the arm and bucket. But Otis realized that additional labor-saving improvements could be made. He moved to Philadelphia and began elaborating on his original design. Within a few years he had convinced the foreman of a locomotive shop to build a prototype, which was immediately put to work building the Western Railroad between Worcester and Springfield, Massachusetts. In 1839, Otis obtained a patent for his steam shovel, but that same year he died from typhoid.

Otis's cousin, railroad contractor Oliver Chapman, put Otis's innovative steam shovels to use. After Otis's death, Chapman married Otis's widow, Elizabeth, and renewed Otis's

> The children's book *Mike Mulligan and His Steam Shovel*, by Massachusetts author Virginia Lee Burton, describes the next disruptive innovation wave, in which the steam shovel is replaced by the shovel driven by the internal combustion engine.

patents. Through his work with John Souther, of Globe Locomotive Works in South Boston, Chapman reinforced and redesigned the original plans. Together, Chapman and Souther began building these shovels in 1859.

The machines were quickly adopted to dig stones and dirt from the suburban Boston hills. The stones and dirt were sent by rail to fill in the tidal flats of the Back Bay, literally creating the basis for one of Boston's most famous neighborhoods. This massive con-

William Otis invented the steam-powered excavator, or "steam shovel," and within five years it was being exported to Russia to help build railroads.

By the date of this image, 1898, the use of steam shovels was widespread. Here, a steam shovel helps with construction of the Wachusett Reservoir in Clinton, Massachusetts.

struction project spanned nearly three decades. Souther steam shovels were also used to build the transcontinental railroad seven years later, as well as a range of other international construction projects. Chapman applied for an improved patent in 1867. During the period between 1839 and 1867, no one else had applied for a shovel patent in the United States, a testament to the unique position of the Chapman-Souther partnership.

It wasn't until the Otis patent ran out that an Ohio company, Bucyrus, began making steam shovels in greater quantity. The early construction of the Panama Canal in the 1880s helped continue the business success of the innovative power excavator for the next fifty years.

1848

Nantucket Sleighride

HOW AN AFRICAN AMERICAN BLACKSMITH GAVE AN EDGE TO YANKEE WHALERS

Few creatures capture the imagination as does the whale: the world's largest animal, majestic, inspiring, a symbol of the beauty and power of ocean life. But in the nineteenth century, the whale was primarily an economic creature. Whale fat, properly refined, was a key lubricant for textile looms and made an outstanding, clean-burning fuel for lamps that lit homes and workplaces around the world. Ambergris, an intestinal secretion in the whale, was used for perfume. And whalebone was needed for the manufacture of corsets and umbrellas.

Pursuing and killing whales could be lucrative, but it was also dangerous. As early as the fourteenth century, Basque ships hunted whales, and for the next six centuries, other Europeans pursued whales in the far reaches of the North Atlantic, all making the final pursuit from small boats called whaleboats and killing with an iron-tipped lance called a harpoon.

Americans, primarily from New England and especially from Massachusetts, began to participate in the business by the eighteenth century. With few resources at home, many in Massachusetts had turned to the sea to make a living. By the nineteenth century, the abundance of whales off the New England coast had created a robust class of entrepreneurs, captains, and crew members who came to dominate this global business, with two Massachusetts towns, Nantucket and New Bedford, assuming preeminence in sending square-rigged whaling ships around the world.

Whale hunting was a mainstay of New England port economies in the eighteenth and nineteenth centuries. Whale oil was critical for fueling lamps and lubricating machinery prior to the development of petroleum alternatives in the 1850s.

In part because of the persistent need to find crews willing to endure the long voyages, low pay, and poor working conditions, New Bedford had acquired a diverse, polyglot population that included many Azorean Portuguese and perhaps even some South Sea islanders like the fictional Queequeg in Herman Melville's *Moby Dick*. This diversity and the connections that a seaport provides to ships traveling along the coast may have been why New Bedford also became the destination of a number of enslaved people who had escaped from the Southern states, including the famous author and abolitionist Frederick Douglass and the accomplished African American blacksmith Lewis Temple.

The harpoons of the early nineteenth century could wound but not hold an animal. They often lost their grip as a whale struggled to escape. Lewis Temple crafted a harpoon tip that changed the industry. His 1848 innovation, which became known as the Temple Toggle Iron, was crafted from two pieces of iron deftly connected with a pivot point. The toggle acted as a single sharp tip initially, but after the harpoon was thrust into the whale, the tip would pivot to one side, making two barbs that extended wider than the entry point. This locked the harpoon to the hunted animal, preventing its escape and often leading to what whalers called the Nantucket sleighride, when the whaleboat and its crew were pulled by the terrified animal until it grew exhausted or died from its wounds.

The concept caught on quickly and greatly boosted the success of whale hunting while reducing danger to the hunters. Later in the

A New Bedford, Massachusetts, statue of Lewis Temple, who perfected the toggling tip harpoon.

nineteenth century, after the industry had begun to decline (the last US whaling voyage occurred in the early twentieth century), a powerful harpoon cannon was invented in Norway that made hunting the whale an entirely one-sided affair. This technology was so successful that some species of whales were nearly hunted to extinction by those who continued to hunt well into the twentieth century. A few countries, such as Iceland and Japan, still use this cannon to hunt whales.

Unfortunately for Temple, he did not patent his design. Although his design earned him fame, he made no real money from his invention. He died a few years later from injuries sustained in a fall. His achievement, however, is memorialized by a statue in New Bedford, Massachusetts.

1850s Clippers
FASTEST SAILING SHIPS EVER

When gold was discovered in the hills of California in 1849, the news spread as fast as the ships of the time could carry it. Gold fever sent people flocking to California by the thousands, seeking to strike it rich. Others saw a different kind of fortune in these fortune seekers, called the forty-niners; merchants understood that there was good money to be made in supplying the forty-niners, so the port of San Francisco became a bustling boomtown. In the first four months of 1848, only four ships sailed from the East Coast for San Francisco; in 1849, 774 sailed for the same destination. The problem was that the ships couldn't get there fast enough. The sixteen-thousand-nautical-mile journey around Cape Horn, the southernmost tip of South America, and up to San Francisco took an average of two hundred days.

Donald McKay, a ship designer, would cut that journey in half. In 1849, McKay launched his first clipper ship, *Stag Hound*, from his East Boston shipyard. A year later he launched *Flying Cloud*, an innovatively designed sailing ship that cut an even sleeker line through the waves. McKay's clippers revolutionized the industry and filled the sky with acres of sail. Their sailing feats fit America's and the world's fascination with speed; they remain some of the fastest oceangoing sailing vessels ever built. Each ship was an embodiment of McKay's search for the perfect seagoing vessel, and he became as famous for innovation as Steve Jobs would be in the 2010s. The Boston

Donald McKay had already established a strong record as a shipbuilder when he was encouraged to set up his own yard in East Boston by Enoch Train, a financier in the shipping industry.

The Boston-built *Flying Cloud*, one of the most famous of the clipper ships, set the world's sailing record for the fastest passage between New York and San Francisco—89 days 8 hours—a record that stood until 1989, when *Thursday's Child*, an ultralight sloop built of the latest high-tech materials, finally made a faster run.

clipper solved the dilemmas that had bedeviled shipbuilders for centuries: it combined great speed with sufficient cargo-carrying capacity and thrived in heavy seas. Although their age was a short one, the clippers were instrumental in opening new trade routes and connecting the world's people through ports of call.

McKay's family had fled the United States during the American Revolution; fifty years later, Donald McKay arrived in New York as an immigrant. Becoming a shipbuilder at a young age, he left the competitive New York City market and went to Newburyport, Massachusetts, where he built his own successful shipyard. The Bay State shipping financier Enoch Train lured McKay to Boston by funding new ships that paid for a new East Boston shipyard. As it has been for so many of the city's innovations, Boston finance, and Train's in particular, was McKay's ticket to success. McKay started working on Train's Boston-Liverpool line of ships. Train was looking for speed, strength, and stability at sea, as well as supreme comfort for passengers. McKay gave it to him.

McKay's invention of the Boston clipper created a race among rival shipbuilding houses in Boston, New York, Philadelphia, New Orleans, Baltimore, and Portland, Maine, to build the world's fastest, most economical ship. But McKay's design could not be beat. In the summer of 1851, his *Flying Cloud* raced the New York–built *Challenger* to see which could get to San Francisco the fastest from a starting point in New York City. On the way south to Rio, they set speed standards by a sailing ship that have not been broken in a century and a half since. *Challenger* was captained by "Bully" Waterman, the *Flying Cloud* by Josiah Perkins Cressy. (Cressy's wife was the navigator, a first for women on clipper ships.) Waterman, who was a sadist and would later be convicted for murder, chained some of his sailors to the rigging during the winter as they rounded Cape Horn, and some froze to death. *Flying Cloud* made it to San Francisco in eighty-nine days; the *Challenger* took 110. Cressy and his wife were raised on the shoulders of well-wishers when they arrived in San Francisco, and Waterman was arrested for murder.

The clipper age was a transitional time in US history. Precariously set between the rapid move west and the Civil War, the age was short-lived. Too many ships, the coming of steam, and the economic Panic of 1857 sent the Boston shipping industry into rapid decline. Geography benefited cities further west with better river and railroad connections to the expanding nation. Boston and New England turned increasingly toward industrial development, investing in factories and machinery more than ships and spars, and pursuing new technologies such as the telephone. However, the region continued to produce ships of all kinds into the twentieth century.

1897
Down Under and Up Above
USING ELECTRICITY TO CREATE AMERICA'S FIRST SUBWAY

The world's first subway, known as the London Underground, was built in that city in 1863. It relied on steam engines, so the system was smoky, and expensive. Later in the nineteenth century, unprecedented advances in electricity technology made underground public transportation a far more appealing way to move people within a city. The city of Boston, because it was one of the hubs of the new electricity industry, set in motion several major innovations that combined technology and the concept of a complete system, including both above-ground public transportation and the first underground transportation system outside of London (and a small one in Budapest), which immediately became a blueprint for metropolitan transit worldwide.

Electricity was key. Thomas Edison (who worked in Boston for a number of years) invented the incandescent lamp in 1879, the most famous electricity milestone, but many other milestones were equally important, including the development of power plants and substations to move electricity to factories, homes, and transportation systems. In the late 1880s, scores of major innovations were developed in the Boston area, including generating apparatus; motors; and the means of manipulating, controlling, and transmitting electricity.

Boston used these innovations to become the first, and the largest, city worldwide to

Building underground was a challenge. So, too, was keeping traffic moving on the surface during the disruption. This view shows the vicinity of Boston's Park Street station, as construction moved below Boston Common.

convert thousands of horse-drawn streetcars, as a whole system, to electric power, doubling speed, lowering costs, and eliminating hundreds of thousands of tons of animal waste from city streets. In converting horse power to electricity, the city adopted the latest developments in electrical generation and management and provided the first example of a unified, large-scale generation and distribution system. To do this, Boston engineer Fred Pearson envisioned and created in the early 1890s the innovation of an electric rapid transit system linking suburbs and the city center via surface routes, and later a critical underground section (expanded gradually over the following century) through the busiest and most gridlocked sections of the city.

The project began with the electrification of streetcars. One of Edison's main rivals in the 1880s was an inventor named Elihu Thomson. Though Thomson is barely remembered today, in the minds of businessmen investing in the new electricity industry his work was different, and rivaled Edison's approaches. He and his colleague Edward Houston set up the Thomson-Houston Company in Lynn, Massachusetts, which built power systems that were well ahead of the work of Edison Electric and Westinghouse.

Frank Sprague had showcased the first electric streetcars in Richmond, Virginia, in 1888, demonstrating that a system could work. However, Sprague's system was not powerful enough to succeed in hilly Richmond. In Boston, Thomson-Houston worked with industrialist Henry Whitney to convert Whitney's eight-thousand-horse streetcar empire to electric trolleys—and to demonstrate that trolleys could go up and down hills.

In December 1888, Whitney's West End Street Railway piloted a multimile, aboveground electrified line, which doubled the speed of horse-drawn streetcars, from suburban, hilly Allston to downtown Boston. Within a year, he had converted most of his extensive system to electricity, using the Thomson-Houston Company for the contract. When completed, the system included a network of multiple generating stations, making it the first complete, large-scale, large-capacity electricity supply system, a world-changing innovation. The key to the system was the world's largest single power station in downtown Boston, developed by Pearson, which drove the whole system.

The heart of the Boston system was one main street with four trolley lines. But gridlock was now intensified by the speed and growing popularity of the electric trolleys. To go four blocks in the center of Boston could take an hour. How could the problem be solved? Henry Whitney proposed a subway, but he would not be the one to build it.

Nathan Matthews, Boston mayor from 1891 to 1894, provided the leadership for an innovative solution: tunneling beneath the historic Boston Common to create an underground path for the quiet and efficient new electric streetcars, thus bypassing and reducing the city's traffic. The subway concept was put to referendum and won, despite considerable opposition. Some believed they would feel claustrophobic traveling underground. Others felt that the dead in the historic burial ground should not be disturbed for a mere trolley tunnel. Hundreds of unmapped graves were found and moved by the time the project was complete.

Funding came from the city as well as local entrepreneurs and local banks. Conscious that New York City also wanted to build the first subway in the Americas (and the third in the world to have a system up and running), Mayor Matthews did all he could to push the subway to completion before New York City. The competition was intensified by the fact that the head of the New York City effort was Henry Whitney's brother, William. It was a high-stakes public duel. Despite several worker injuries and deaths—and one disaster involving a gas main explosion—the Boston subway opened on September 1, 1897, immediately clearing gridlock along Tremont and Boylston streets. A hundred thousand customers used it the first day. Within nine years, Boston had created the first fully electric trolley system in the world, and the first subway in the Americas.

In the meantime, key players in New York City had not been able to agree on a plan until after Boston's subway opened. New York City's first line, the Interborough Rapid Transit (IRT), would not open until seven years after the opening of Boston's subway.

First trolley car to arrive by subway underground in Western Hemisphere, at Boston's Park Street station, in September, 1897.

THAT'S ENTERTAINMENT

1891 Slam Dunk

JAMES NAISMITH AND THE INVENTION OF BASKETBALL

In December 1891, Dr. James Naismith had a problem. Naismith was a Canadian immigrant and physical education instructor at the international training school for the Young Men's Christian Association (YMCA) in Springfield, Massachusetts, now known as Springfield College. He wondered how he was going to keep his gym class occupied when the weather was bad. He needed an activity. His boss gave him two weeks to come up with an indoor game that "wouldn't be too rough."

With limited space available and the constraints he was given, Naismith set out to create a game that was energetic but safe. Inspired by elements of traditional children's games, he put pencil to paper and sketched out a plan. That plan became the familiar and globally popular game of basketball.

Initially, the baskets really *were* baskets: bushel baskets made of thin wooden slats and normally used to transport or store peaches and other fruits and vegetables. Naismith nailed two of these to opposite walls, ten feet above the floor. The ball was a soccer ball, and the number of players was formulated to allow a single American football team, unable to pursue outdoor practice, to be divided into two basketball teams.

Naismith was not thinking in terms of anything more than keeping his local student group occupied indoors. However, the game was refined over time and quickly proved to be very popular, particularly in climates where the weather often restricted outdoor play. It spread first among the community of US and

Tasked with coming up with an indoor game to keep athletes active during inclement weather, an inspired James Naismith invented basketball—and the first "basket" was a peach basket.

Boston Celtics star Larry Bird drives for a layup against Magic Johnson and the Los Angeles Lakers in an NBA game at the Boston Garden.

Canadian YMCA facilities and then expanded to colleges and high schools. By the early 1900s, prominent institutions such as the University of Chicago, Columbia University, Cornell University, Dartmouth College, Yale University, and the US Naval Academy all had basketball teams. Naismith himself coached at the University of Kansas for six years. The traditional peach basket lasted until 1906, when it was finally replaced with a metal hoop.

More changes came over the years, including the brighter, orange-tinted ball introduced in the 1950s, but the game of basketball remained remarkably true to Naismith's original plan. Today, basketball is played by people of all ages and on every continent. Not far from where basketball was invented, in a huge facility beside the Connecticut River in Springfield, Massachusetts, the National Basketball Hall of Fame commemorates the unlikely invention that changed the world, its inventor, and the millions who have played and enjoyed basketball for more than 125 years.

1903

Batting Around a New Idea

THE FIRST WORLD SERIES

Major League Baseball's (MLB's) World Series, often called the Fall Classic, is a US institution, the culminating annual event of the country's so-called national pastime. Although today we may think that the World Series has been around forever, its history began in 1903, when two cities—Boston and Pittsburgh—collaborated to create a competition that was truly innovative and truly long-lasting.

Barney Dreyfuss, owner of the National League's Pittsburgh Pirates, and Henry Killilea, owner of the Boston Americans (soon to become the Red Sox), invented this season-ending series of games to determine which league's champion was the best team in baseball for that year. The American League was the new kid on the block, founded in 1901 to replace an earlier league; the National League had been around since 1876. The Boston Americans had only come into existence with the new league; many of its players had played for National League teams before being drafted by the new league in 1901.

But the Boston team was anchored by one of the greatest pitchers in the game, Cy Young, who was already thirty-four years old by the 1903 series. "Cy" was short for "Cyclone," a nickname acknowledging Young's strength and durability. He once pitched all eighteen innings of a doubleheader and won both games, and he would go on to win more professional baseball games (511) than any other pitcher.

The presidents of both leagues encouraged the development of a new postseason series, but the two team owners drove negotiations and actually owned the series that first year. Dreyfuss proposed an eleven-game series, but after discussions with Killilea, he agreed to a best-of-nine format, to begin in early October.

The success of the 1903 series spurred attempts to schedule a regular championship series between the two leagues the following year, but it was 1905 before the National Commission, baseball's governing body, established a World Series format that was accepted by both leagues and would last until the present day.

Permanently locked in concentration, Cy Young's statue sits on the spot where the pitcher's mound for the first World Series was located—now part of the campus of Boston's Northeastern University.

Take me out to the ballgame! The first World Series was played in part on Boston's Huntington Avenue, and at one point fans broke through a rope barrier and took over the outfield.

The 1903 championship provided a great opportunity for Boston to display its civic pride. A booster club called the Royal Rooters, hundreds strong, led the Boston charge. The group traveled by train to attend all the Pittsburgh home games. Led by Mike McGreavey and John "Honey Fitz" Fitzgerald (later mayor of Boston and grandfather of President John F. Kennedy), the Royal Rooters drank heartily, wore colorful clothes, and sang their theme song, "Tessie," with new lyrics targeting the Pirate players.

The Pirates were a strong team, led by the great hitter Honus Wagner. They won the first game, which was played at Boston's Huntington Avenue Baseball Grounds (now part of the campus of Northeastern University), 7–3, with Pittsburgh's Deacon Phillippe outpitching Young. Boston bounced back to win game 2 in Boston, but the Pirates took the third game 4–2. The teams headed to Pittsburgh with the Pirates holding a 2–1 series lead. When the home team won the fourth game, it was widely believed that the Pirates would win the series. But Young's pitching and Boston's potent offense led the American League to victory in the fifth game. Buoyed by this victory, the Boston team would win the next three games—and the first World Series.

The series had been a hit. Boston won the first championship and four more between 1905 and 1918. In 1918, the team traded Babe Ruth to the New York Yankees and had to wait another eighty-six years until its eventual redemptive victory in 2004.

1939 Technicolor
BRINGING LIFE TO THE BIG SCREEN

The cultural and social impact of film in the early twentieth century is difficult to overstate. Audiences were enraptured, absorbed, and transported, even by the earliest black-and-white moving images. But the invention of color—in particular, a process known as Technicolor—took the experience to a whole new level, culminating in cinematography that relied heavily on its brilliant palette to stun and amaze audiences and, more subtly, to affect audiences emotionally and psychologically.

Early attempts, primarily in Europe, to deliver color films had proven problematic and expensive. Then two MIT-trained engineers, Boston natives Herbert Kalmus and Daniel Comstock, and their partner, self-trained engineer W. Burton Wescott, believed they could solve the problem of color. They named their partnership KC&W. It had researched and patented a number of innovations in fields as diverse as abrasives and the reprocessing of slaughterhouse wastes. But the idea of color movies became their passion. With financial support from a Boston investor, they launched the Technicolor Motion Picture Corporation in 1914. While most of their early work was based in Boston, the trio converted an old Pullman railroad car to a rolling laboratory and office to make test films and showcase the result.

Technicolor would eventually become synonymous with blockbuster hits, but its first three versions, which were released from 1917 to the early 1930s, did not have much impact. These films excited audiences, but costs were relatively high, and Hollywood studios saw no reason to move from black-and-white film. In the late 1920s, however, another new technology, sound recording, introduced moviegoers to "talkies." This experiment turned into a seismic shift as audiences rejected silent films and often their stars. Within a few years nearly all films featured voices and music.

Into this environment, Kalmus and his partners introduced a fourth version of Technicolor, with the largest palette of color yet offered by any film system. Unlike the earlier versions, which used *alternate frames* of the film to capture different parts of the spectrum, this system used *three*

Technicolor brought blazingly bright color images to moviegoers in the middle of the twentieth century. Different reels of film captured different parts of the visible spectrum, which projectors then combined for theater audiences.

separate strips of black-and-white film, each of which recorded a different primary color (blue, red, or yellow). When recombined and projected on the screen, the results were stunning. Although few studios were interested at first, a new animator named Walt Disney and his brother, Roy, decided to spend the extra money and switch their short feature, *Flowers and Trees*, to the Technicolor process. The result was a smash hit and an Academy Award for Disney in 1932.

The Great Depression continued to deepen, however, and the success with Disney did not immediately translate into a booming business. That success had to wait until 1939, when two historic smash hits were filmed in Technicolor: the Civil War epic *Gone with the Wind*, and the fantasy story *The Wizard of Oz*. From then on, as with talkies a decade earlier, audiences demanded better-than-life, dazzling color. At last, "Filmed in Technicolor" became almost as important as an all-star cast in ensuring a movie's financial success.

By the mid-1950s, however, Technicolor's heyday was over. First, the disruption of World War II slowed the adoption of Technicolor. Then, in a few short years after the war, television went from an experimental technology to a mass-market phenomenon, cutting deeply into the profits of the movie business and making studios more cost conscious. Additional advances in film chemistry allowed competitors such as Kodak to begin to offer full color (although not as brilliant as Technicolor) using a single strip of film.

The technical achievements in color reproduction remain admired today, and frequent

The Technicolor breakthrough made color movies compelling and special.

attempts have been made to revive aspects of the process. Focusing on other aspects of the film business, Technicolor Motion Picture Corporation remained an important player in the industry. In 2001, Technicolor became part of the French electronics and media conglomerate Thomson, and in 2010 the entire company was renamed Technicolor SA, after its US film technology subsidiary.

2007

The Spark That Lit Kindle

MAKING ELECTRONIC BOOKS A REALITY

In an era when more and more information is consumed via computer screen, the invention of the electronic book was almost inevitable. However, it took a group of Boston-area researchers to develop the process that led to the Amazon Kindle, the start of a revolution in publishing. Amazon released its first e-reader in 2007. It sold out in hours. Today, many books appear in electronic form, and Kindle has been followed by Barnes & Noble's Nook, and other e-readers.

The Greater Boston area has a long history of leadership in publishing technology and practice, beginning with the first printing press and the first newspaper in British North America. Thus, it's not surprising that electronic ink (e-ink), the critical technology behind not just the Kindle but many wider applications, was developed here.

Two features distinguish e-ink. First, like type on a piece of paper, e-ink is intended to be easy to read from any angle and in a wide range of lighting conditions (it depends on external lighting). Second, and just as important, because it uses far less power than most other forms of electronic display, it can display characters and images for a long time—an especially helpful feature in a portable reader device.

Creating these features in a reliable and affordable package was the effort of a large team: Dr. Joseph Jacobson at MIT, who first imagined how wonderful it would be to have a "book" that could hold hundreds of titles; two MIT undergrads, Barrett Comiskey and J. D. Albert, who contributed to developing the technology; institutional involvement from the MIT Media Lab, along with help from Lexis-Nexis founder Jerome Rubin; and Russell Wilcox, with his strong start-up experience, who was recruited as CEO, then created the business plan and helped attract support from local venture capital firms to help the company get off the ground. With Amazon, Wilcox and the e-ink team developed the Kindle.

All electronic paper systems rely on changing electrical forces to shift the appearance of a colorant in the "paper" material. In the case of e-ink, the basic approach used microscopic spheres of white floating in a dark liquid.

Mimicking some of the most useful properties of paper—permanence and erasability, for example—
E Ink Corporation created the underlying technologies used in a wide variety of display technologies.

Changing the electrical charge across each of the small cells within the thin plastic electronic paper, or display screen, causes some part of the screen to appear black and other parts white, depending on which direction the electrical charge has moved the spheres. The resulting high contrast makes for easy reading and a small amount of power needed to change or sustain the display—ideal for the human pace of reading, even if there are interruptions. E-ink's color displays use the same principle, although the action is a lot more complex. In addition to making e-readers possible, e-ink's technology has also found application in other electronic devices, signage, and more. Here again, it took multiple factors to get a good idea to market and broad commercial acceptance. Further innovations have now extended its reach and made it pervasive.

CONNECTING PEOPLE

1690 *Publick Occurrences*
AMERICA'S FIRST NEWSPAPER

In 1690, European settlement in New England was still little more than a hundred miles wide. Benjamin Harris, a recent émigré to Boston from London, began publishing *Publick Occurrences, Both Forreign and Domestick*, with the help of printer Richard Pierce. Started with lofty goals and the hope of serving the public's thirst for information, it was the first newspaper in all of British North America.

Although *Publick Occurrences* measured only six by ten inches and contained some blank pages, it aimed to deliver real news on a monthly basis. Harris himself had already been publishing books, pamphlets, and newspapers in London, but his controversial views had got him in trouble with authorities. Upon arriving in Boston in 1686, he established the London Coffee House, a place to meet and peruse imported periodicals. The coffee house was even open to women, which was exceptional for that time. Despite its lofty goals or perhaps because of them, *Publick Occurrences* survived just four days: local authorities found its content, in particular its mild criticism of the government, to be unacceptable. Furthermore, it was published without a crown license.

Harris stayed in Boston for a few more years and continued to publish almanacs and even books for the government, but he did not publish another newspaper until after his return to London in 1695. His departure was not the end of the story, however. Other Bostonians were well prepared to follow his lead. As early as 1638, the first printing press in America was brought to the colony from England by a clergyman, who died on the voyage. But his wife, the skilled printers he brought with him, and the press survived and were promptly set up in Cambridge at Harvard College. That print shop was put to work publishing almanacs and books of psalms. Other printers followed, creating a small-scale printing and publishing industry in the colony. All of this printing and the strong focus on education in the colony made for a highly literate population hungering for what a newspaper could provide.

Fourteen years after *Publick Occurrences*, Bostonians tried again with the *Boston News-Letter*, which became the first continuously published newspaper in British North America. It was operated in cooperation with the government, however, and published only what the government deemed fit. With that comfortable arrangement, it persisted through several owners and editors until 1776, when the British withdrew from the colony.

While true freedom of the press, expressed most forcefully in Article I of the Constitution, would take much longer to evolve, other Bostonians took important steps. In 1721, for instance, James Franklin began publication of *The New-England Courant*, noted for its high-quality essays and frequent irreverence with regard to local authorities and local practices. James's not-yet-famous younger brother, Benjamin Franklin, worked as a typesetter and wrote a series of articles under the pen name of Silence Dogood, one of which led to the imprisonment of his brother. Such incidents eventually led to the suppression of the publication by the government. By that time, Benjamin had relocated to Philadelphia.

Boston's "news media" continued to evolve. It played an important role in the region's political development and eventually its rift with Great Britain in 1775. Continuing to the present, Boston has maintained a lively publishing and news industry. The city is one of the few of its size in the United States to still support two daily newspapers: the *Boston Herald* and the *Boston Globe*.

The first newspaper in what would become the United States, unsanctioned by government, was published in Boston . . . and promptly banned by authorities.

1876
Bell's Telephone
FROM EXPERIMENT TO REALITY

Can any of us imagine life without a phone? In the digital age, the telephone is a constant and must rank as one of the world's most important innovations. And most of us know that the inventor of the telephone was the Scottish-born Alexander Graham Bell. But how many of us know that Bell's invention was developed in Boston in 1876? In the early 1870s, a number of inventors in the United States and Europe were working to develop something like the telephone. Why and how did this materialize in Boston rather than Edinburgh, Chicago, or Paris?

After the Civil War, Boston was a world center for new ideas in science, invention, and social change, and Bell eventually came to live and work in this hotbed of innovation. He was the son of Alexander Melville Bell, a Scottish innovator who had started a movement to help the deaf by creating a system of what he called visible speech based on phonetic symbols. Discouraged about prospects for the system's further adoption in Europe, father and son left the United Kingdom for Canada, where Alexander began his early work on his telephone.

Canada did not offer the Bells many opportunities to work with the deaf, but Boston did have a need for the visual speech system. Bell's father decided not to go, saying he was too old, but Alexander made the move, establishing a school in Boston to train teachers of the deaf in 1872. The school was part of Boston University, where Bell was appointed professor of vocal physiology in 1873. But he also continued to work on the telephone, soon discovering that his new city had the right resources to support his inventiveness.

Alexander was excited and motivated by Boston's intellectual and political activity. A day or two after his arrival, he noted in his journal that he heard one of the first black congressional representatives, elected during Reconstruction, speaking out against the Ku Klux Klan. Within a few weeks, he had befriended many prominent educators, including Professor Edward C. Pickering of MIT and Lewis Monroe, dean of Boston University's School of Oratory, both of whom would help Bell develop his new invention. Pickering, in particular, performed elaborate scientific replications of the newest experiments, and much of the early investigation was done in Pickering's labs.

Scotsman Alexander Graham Bell found Boston to be fertile ground for pursuing his dream of a voice transmission instrument. The city had a community of inventors and skilled "mechanics" able to turn ideas into workable machines. It also had investors, including the man who became his father-in-law, willing to consider supporting new ideas.

Building on this foundation, Bell expanded, renting in a building that several years before had been used for its lab space by Thomas Edison, the other major American inventor of the late nineteenth century. Through teaching deaf students full-time, Bell met his major financier and key supporter, Gardiner Green Hubbard, the father of his wife-to-be, Mabel Hubbard, who was one of Bell's deaf students. The funding Hubbard provided, along with the network of contacts Bell had established after his move to Boston, would be crucial in moving his idea from concept to actual, working invention—a great example of the value of bumping and connecting in the fertile environment of Boston.

Bell's breakthrough came early in 1876, and by the time of the national exposition in Philadelphia that summer, he was able to demonstrate it in public to scientific, business, and world leaders. Within three months, he had hired the telegraph company on the first floor of his Boston lab building to build the first telephones and hook them up. The new Bell Telephone Company anchored its national headquarters on Pearl Street, ten blocks away, developed telephone lines and exchanges, and gradually consolidated the strongest position in this new industry. The Bell telephone system, which morphed into AT&T, was based in Boston for forty years before the company moved to New York City.

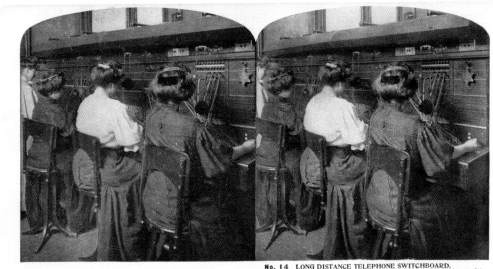

No. 14 LONG DISTANCE TELEPHONE SWITCHBOARD.
SEARS, ROEBUCK & CO., Chicago, Ill.

The invention of the telephone quickly spawned the engineering needed to connect telephones into a useful network. Operators—almost invariably women—accomplished the switching and interconnecting and answered customer questions.

1969

Connecting the World

THE TECHNOLOGY THAT POWERED THE BIRTH OF THE INTERNET

What did people do before the internet? Its effect, from the way it moves information and connects people and human activities instantaneously and affordably, has been so profound and on such a large scale, that it's hard to remember when it wasn't part of our lives. Many of us have now grown up using it.

The internet might not have emerged so quickly and so fully but for a fortuitous investment by the US Department of Defense (DoD) and its Advanced Research Projects Agency (ARPA), an investment that built on the work of Boston-area researchers, particularly those at a small company called Bolt, Beranek and Newman (BBN). That research later resulted in the foundation for the World Wide Web: the Advanced Research Projects Agency Network (ARPANET), an early packet-switching network that formed part of the foundation for the internet.

In the 1960s, electronic communications, including television broadcasting, telephone calls, and the occasional computer link, were conducted across millions of miles of copper wire, and every connection had its own wire connections from start to finish. Such a network worked well, but it was expensive to build and maintain.

At the same time, the computing industry was in a period of ferment. At MIT, theorists like Claude Shannon had shown how all communications and indeed all information could be reduced to a collection of bits (binary digits). And pragmatists, like the former Harvard and MIT technologist J. C. R. Licklider, one of the BBN researchers, gave the first public demonstration of time-sharing, a process allowing multiple users to share a single computer. Time-sharing was a vital step forward when computers were expensive and few in number.

In the late 1960s, there were several scientists and firms working on developing an innovative system of moving small packets of information at the speed of light. This Interface Message Processor (IMP) would allow far more communication instances to share a single physical circuit. It also allowed components of a message to reach the recipient through multiple routes, in effect creating an adaptable and survivable network. This new way of transferring information between computers at major universities was critical because telephone lines didn't have the bandwidth to transfer information, and telephone line failures were still quite common in the 1960s.

During this same period, Licklider became the head of the Information Processing Techniques Office at ARPA. His experience with computers and communications, as well as his awareness of other emerging technologies, enabled him to guide the discussions that would lead to DoD funding for ARPANET.

When ARPA's proposal for IMPs arrived at BBN, the company recognized the potential that such a concept could have on the future.

A number of large companies, including Raytheon, had competed for the contract, but BBN shocked the electronics industry by winning the bid. Frank Heart was put in charge of the project to design the IMPs utilizing underlying computer hardware made a few miles away by Honeywell.

Upon receiving the ARPA contract in the final days of 1968, BBN had just nine months to deliver the first IMP to the initial test sites. After many agonizing late nights, the first two IMPs were delivered to the University of California, Los Angeles (UCLA) and Stanford University in September 1969, with Heart sending along BBN employees to chaperone the two machines, located 350 miles apart. Although some of the colleges were initially reluctant to participate, everything worked according to plan. By March 1970, BBN had linked their own machine in Cambridge to four IMPs out west, creating the first cross-country circuit for ARPANET. The network began to expand almost immediately, with new nodes established at the rate of about one a month and with BBN overseeing operations.

A young software engineer at BBN, Ray Tomlinson, developed an electronic software program for communicating between nodes. While setting it up, he needed something to identify which computer the sender was from so that he could easily return a message. He looked at his keyboard to find a symbol, and his eye focused on the little used @ key. He added it to the BBN program. One of the early e-mails he sent was a lunch invitation at a local Chinese restaurant to a coworker.

In October 1972, ARPANET was demonstrated publicly at the International Conference on Computer Communication in Washington, DC. With the new network in place, novel ways to use it, such as email, discussion groups, and gaming, soon caught on. What had begun as a means of ensuring the ability of university researchers to communicate their research data with other researchers had begun to evolve into what we know today: an almost infinitely expandable network where one can turn for information, amusement, community, and personal expression.

Over the years, IMPs were developed into gateways, and the scale of the internet became vastly larger. The ARPANET became the large-scale testing ground for the theory of moving data in packets, which has since become key to the internet, the Web, and even our phones and media delivery systems.

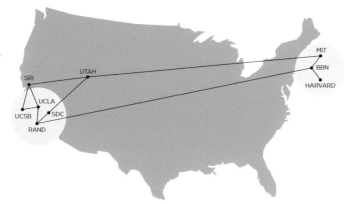

Engineered and managed from Cambridge, the early connections of the ARPANET demonstrated for the first time the practicality of a large-scale digital communication network and served as the starting point for creating the World Wide Web.

1979

The First Killer App

THE ELECTRONIC SPREADSHEET

In the late 1970s, computers and the internet were confined to large businesses, government agencies, and universities that had the resources to invest in these new technologies. A handful of companies, including Apple, were busy trying to popularize personal computers (PCs) cheap enough for average consumers to purchase for home use. But with prices for a reasonably equipped PC running at well over two thousand dollars, most consumers wondered why they would spend that kind of money on a PC when it couldn't do anything that a top-of-the-line, and cheaper, International Business Machines Corporation (IBM) typewriter could do.

The electronic spreadsheet, an invention by two computer experts who had lived for a long time in the Boston area, played a key role in changing the PC landscape and, within a single year, propelled the PC onto hundreds of thousands of desks in homes and businesses. The time-consuming task of preparing paper financial spreadsheets by manually filling in and correcting rows and columns of data was dramatically transformed by Dan Bricklin and Bob Frankston in an attic apartment in the Boston suburb of Arlington in 1979. A Harvard MBA student, Bricklin daydreamed in class about a better way of completing spreadsheets for class assignment than rewriting figures on paper and totaling the results with an electronic calculator, then the hot new tool in business and academia.

Prior to Harvard, Bricklin had worked for several years at Digital Equipment Corporation (DEC), helping perfect word-processing systems for newspaper and publishing companies. Couldn't something similar revolutionize accounting? Over a fall weekend, Bricklin wrote the first draft of what would later become VisiCalc, a spreadsheet software program that would fundamentally alter both the financial and computer industries. His Harvard Business School professor asked a what-if question that required in-class calculation. Using his software, Bricklin had the answer long before anyone else.

Bricklin knew that the key to VisiCalc's success was to make it a product rather than a program. And that required an intuitive user interface. Consulting with an informal group of professors and students, Bricklin developed an interface that the software publisher Dan Fylstra said made VisiCalc "come alive visually. . . . In minutes people who

Dan Bricklin and Bob Frankston, creators of VisiCalc, the first electronic spreadsheet software.

have never used a computer are writing and using programs." By the fall of 1979, VisiCalc was ready to be released. It did not take long for people to realize that they *needed* VisiCalc; in just the first two weeks after the software's release, close to 1,300 units had been sold. The first ad asked: "VisiCalc, How did you ever do without it?"

The quick success had something to do with Boston's close-knit computer community. The Boston Computer Society included many innovators, and when one came up with something, the other six hundred members let each other know. Such local networking can be crucial to an innovation getting traction. And with the Boston forum and his informal network getting him started, Bricklin was ready for his next big step: getting in touch with Steve Jobs.

Bricklin liked the user interface for the Apple II and, because he had limited resources, decided not to develop versions for competing brands. Thus, only customers who purchased an Apple II could use the only electronic spreadsheet on the market. "That's what really drove the Apple II to the success it achieved," Jobs would say later. Sales of VisiCalc quickly reached a thousand units a month, and they soon tracked Apple II sales. About one copy of VisiCalc was sold for every three Apple IIs sold, according to Bricklin.

VisiCalc took the financial industry by storm. Its flexibility and simplicity gave users more options and allowed for additional fiscal models to be drawn up based on different variables and factors. It was the first so-called killer app: an application so powerful that customers were willing to buy hardware or equipment just to be able to use it. Customers were able to justify the cost of a new Apple II because the purchase would pay for itself in a matter of weeks, if not days. The

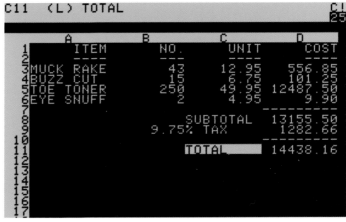

It may look antiquated now, but in its time VisiCalc was the first killer app.

time saved (and stress avoided) using VisiCalc saved money, a concept easily understood by financial analysts using the software.

So why didn't Bricklin become wealthy like Steve Jobs and Bill Gates? Software applications could not be copyrighted in 1979. The US Copyright Office and Congress, and court decisions such as *Apple v. Franklin* in 1983, gradually enhanced the protection for software creators, but it was just too late for Bricklin. Although the company expanded to offer products for other popular computers such as Atari, TRS-80 Model I, Commodore PET, and even the IBM PC, competitors such as Lotus Software, with its Lotus 1-2-3, and much later Microsoft's Excel, soon eclipsed VisiCalc. Bricklin, however, garnered an honored place in the pantheon of innovators.

2001 Ask Them

HOW COLLECTING OPINIONS FROM REAL PEOPLE REVOLUTIONIZED TRAVEL AND SHOPPING

Learning about travel destinations was once done only by word of mouth or by reading guides and travel articles, typically written by a single person. At best, you would get just a few opinions about places you might want to visit and about places you could stay while there. But within a single year in the early 2000s, a small start-up changed the way millions of travelers learned about possible destinations.

From their office above a pizza parlor in suburban Boston, several entrepreneurs tried to determine why they were failing to grow a business-to-business travel site. Steve Kaufer, inspired by a business idea from his wife, had teamed up with cofounders Nick Shanny and Langley Steinert in 2000 to launch a company that would curate syndicated travel content, including expert reviews and online newspaper articles, to the early major travel websites like Expedia and Travelocity and to help travelers plan their trips.

Raising $1.1 million, half from venture capital funds and half from family and friends, they started the company. But sales were low, and as the reserve of company funds dwindled, Kaufer and others urged employees to come up with innovative ideas to pivot the business toward something more lucrative. At one session in their small office, a staffer suggested that they put a little button on the website asking visitors to "write a review." They tried the idea.

Kaufer thought that only complainers with war stories, for example, about a terrible night at a hotel, would contribute. "Who else would want to take the time to write unpaid about their own experience?" he reasoned. To the team's surprise, more reviews than anticipated, ranging from positive to negative, were posted in the first month. They put these reviews on the website, and before long the revenue from the clicks on the ads on their website outnumbered syndicated-generated revenue by a hundred to one. Clearly, user-generated feedback trumped their original business-to-business concept.

The team's new consumer-centric approach became a disruptive innovation that changed the way millions of travelers learned about the places they wanted to visit. The key was to put user-generated content and ratings about hotels, restaurants, and attractions on a website so that the public could access information and contribute reviews and pictures.

Calling the new website Tripadvisor, the business created an arena of unprecedented transparency on a global scale. Instead of the insights of one expert travel writer, the site brought the perspectives and insights of millions of consumer reviewers online, often supplemented by contributed images that provided additional clarity. The reviews, ratings, and photos on the site had an additional, originally unforeseen impact: hundreds of thousands of hotels, restaurants, and ancillary businesses now had the incentive to improve their guest experiences because of the user feedback. Transparency changed how and why travelers made their plans.

The entrepreneurs who launched Tripadvisor didn't start with the idea that they would change the travel industry. They only wanted to improve it in a small way. When user-generated reviews started, Tripadvisor asked for numerical ratings. Amazon was the only other site that had begun soliciting user reviews (about books), and it was not common in 2001 for corporate websites to encourage expanding on reviews with ratings. By gradually decreasing the value of older reviews in its traveler-ranked algorithm, Tripadvisor gave a dynamism to the rankings. Hotels and restaurants had a clear incentive to learn from critical reviews and use them to discover ways to improve.

Tripadvisor's test-and-learn form of innovation led to a spectacular increase in the number of reviewers and site users. The

Tripadvisor used the power of crowds to create a widely consulted and trusted source of travel information.

website currently has well over a half billion reviews. And the innovation in transparency has quickly spread globally, putting consumers in a position of influence. Tripadvisor has had a hugely disruptive but positive impact on the industry that its users report on. Pretty good for a business that was very close to closing down after its first year.

2004

A Social Network

HOW FACEBOOK CONNECTED THE WORLD

A simple idea, conceived in a Harvard University dorm room, changed the world. That's the Facebook phenomenon. It was originally created to provide a campus-wide source of information for and about students, and it was named after the paper-based facebooks traditionally distributed in many colleges. Facebook was launched on the Harvard campus in the winter of 2004. In its first twenty-four hours, 1,200 students had signed on; within a month, half of the student body had joined. The new network spread to other colleges near Boston, then nationally. In 2006, Facebook opened its pages to anyone who met an age requirement (thirteen years old). Now it has over 2 billion monthly users: *2 billion*, about 30 percent of the world's population.

The CEO of Facebook, Mark Zuckerberg, began programming when he was in middle

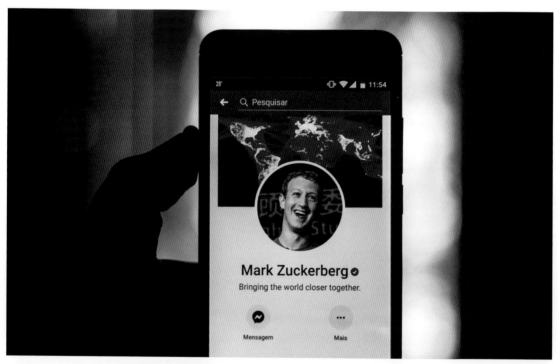

Mark Zuckerberg conceived of Facebook while still an undergraduate at Harvard. He dropped out to build the company but received an honorary degree from Harvard in 2017.

school. He developed an application for his father's dental practice. Zuckerberg's creation of Facebook was so successful that he dropped out of Harvard to bring it to market (the university would grant him an honorary degree in 2017). A few months after launching the college version, Zuckerberg and his roommate Eduardo Saverin met with a representative from a Boston-based venture capital firm, Battery Ventures, who decided not to invest in the start-up because the firm thought the concept was a fad, and the representative had recently invested in one that had failed! But when Zuckerberg and four partners moved the company to Silicon Valley in June 2004, they secured funding from the venture capitalist Peter Thiel, a cofounder of the online payment system PayPal.

The company's growth was staggering. From simply sharing images and basic information about members, Facebook went on to become a platform for chatting, debates, fundraising, advocacy, commercial promotion, and more. It provided a common environment for families to share and stay in touch, for meeting new people, and for sharing images and videos. The company went public in February 2012, valued at $104 billion, the largest valuation for a newly listed public company up to that point. It reached a billion users that same year.

Facebook has profoundly changed the way people communicate and interact, personally and politically. It has had a huge impact on the delivery of news and influenced how businesses interact with customers. With a good deal of controversy, it has contributed to making data privacy a global concern. By any definition it is a world-changing innovation, and we have yet to assess its full impact.

These days, Facebook points to its impact through several key metrics, including more than $2 billion raised through the platform by members in support of the causes they care about, 100 billion messages transmitted daily, and more than 140 million businesses showcasing themselves online.

In terms of broader cultural impact, Facebook and Zuckerberg have been the subject of films (*The Social Network*) as well as appearances and references in books, music, the political world, and the web.

Collins English Dictionary once declared "Facebook" as its new word of the year, and the verbs "friend" and "unfriend" have become part of the language. Zuckerberg and his wife, Priscilla Chan, have joined fellow billionaires Bill Gates and Warren Buffett in making "The Giving Pledge," which commits them to donate at least half of their wealth over time.

BRAVE NEW WORLD

1926 Rocket Man
DREAMING OF SPACE FLIGHT ON A MASSACHUSETTS FARM

Robert Goddard was a young dreamer intrigued by the science fiction of H. G. Wells's novel *The War of the Worlds*. He experienced an epiphany when he climbed a cherry tree and became absorbed by his view of the sky. As he wrote later, "I looked toward the fields at the east, I imagined how wonderful it would be to make some device which had even the possibility of ascending to Mars, and how it would look on a small scale, if sent up from the meadow at my feet. . . . I was a different boy when I descended the tree. . . . Existence at last seemed very purposive." Goddard's pursuit of the vision he had that day would indeed help humanity travel in space.

Although he suffered health problems as an adolescent, Goddard was an avid reader, not just of fiction but also of complex works such as Isaac Newton's *Principia Mathematica*. He found that Newton's third law of motion (an action by a body creates an equal and opposite reaction) offered theoretical underpinnings for the idea of space travel. Earning a degree in physics from Worcester Polytechnic Institute, Goddard undertook graduate studies at Clark University, also in Worcester, Massachusetts, from which he received a master's degree in physics in 1910 and a PhD in 1911. Staying on at Clark University as a research associate, Goddard quietly plunged into study of solid-fuel rockets, attracting support from the university and even the Smithsonian Institution.

Goddard's work showed that the efficiency of rockets and the potential for space travel could be dramatically increased by adding a nozzle to further accelerate the hot gases. He also made pioneering experiments on ion thrusters—a concept today used to propel craft in deep space. During World War I, he

The "giant step" American Neil Armstrong took on the moon in 1969 was enabled by Robert Goddard, who crafted the theoretical basis for rocket flight. He tested and proved his ideas in a short 1926 test flight from a field in central Massachusetts.

worked for the US Army on a simple rocket that could be fired from a tube, work that his colleagues on the project later applied to the bazooka used during World War II. The bazooka helped infantry soldiers fight tanks and attack fortified positions.

But Goddard's introduction to the wider public didn't come until after World War I, and it wasn't pretty. In 1919, he published a scientific paper, "A Method of Reaching Extreme Altitudes," now considered one of the seminal texts in the field of rocketry and space travel. This provocative and highly influential paper demonstrated the theoretical possibility of reaching space, entering orbit, and even traveling to the moon by means of powerful rockets. His paper went so far as to outline the size of such rockets and the amount of fuel they would use.

Unfortunately, the press found even this theoretical suggestion of a humanmade craft embarking for the moon to be laughable. In a front-page story, the *New York Times* portrayed Goddard as a crackpot, insisting, as did some other scientists, that Newton's third law would not allow a rocket to function in space! The incident scarred Goddard and made him far less willing to share his later research. Nevertheless, he persisted in his studies.

The most important fruit of his research in these early years was his recognition that liquid-fuel rockets would be critical for space travel and helpful for many larger rocket designs. But Goddard was never satisfied with being merely a theoretician. In 1926 he built and successfully launched the world's first liquid-fuel rocket from his aunt's farm in Auburn, Massachusetts, an achievement he did not publicize.

In 2003, to honor and commemorate the work of Robert Goddard, engineers at the Marshall Space Flight Center built and launched an accurate replica of his first rocket to take flight.

Why liquid fuel? First, in most cases the most powerful rockets can be made using only very dense liquid fuels combined with equally dense oxidizers (typically, liquid oxygen). Thus, it is no surprise that most rockets bound for space use liquid fuel. Just as important, however, is the control offered by liquid fuel. Solid-fuel rockets have existed for perhaps a thousand years, first created by the Chinese using gunpowder, and they are comparatively inexpensive and generally reliable. Second, solid-fuel rockets are not controllable once they begin to burn because the process must continue until the fuel is gone. Liquid-fuel rockets, on the other hand, can be stopped and restarted at will, which is ideal for space travel or even for the reuse of rockets.

After 1926, Goddard focused single-mindedly on perfecting his invention, with the goal of eventually opening the door to space travel. Funding from Clark University and later from the Guggenheim Foundation was limited, and ill health also slowed Goddard's success. Goddard nonetheless demonstrated or tested a variety of innovative rocket features, such as turbopumps for fuel, steerable rocket motors, and the use of the liquid fuel itself as a coolant to extend the life of motors. He influenced Theodore von Kármán, the founder of the Jet Propulsion Laboratory (JPL) at California Institute of Technology (Caltech), and, of course, Wernher von Braun, who built the infamous V-2 rocket for Hitler's Germany. After the war, von Braun and other German scientists went to work for the US military and then the National Aeronautics and Space Administration (NASA). The ultimate result was the liquid-fuel American Saturn V rocket, more than 360 feet tall, which propelled the first humans to the moon in 1969.

Goddard died in 1945, before he could see the triumph of his concepts. However, he is memorialized in numerous ways, including NASA's Goddard Space Flight Center. Some of his harshest critics, most notably the *New York Times*, ultimately changed their tune. The *Times* printed a correction at the time of the moon flight, acknowledging their error regarding Newton's third law of motion and its implications for Goddard's theories.

1953 Count on It
GIVING EARLY COMPUTERS A MEMORY

By the mid-1940s, just after World War II, scientists and engineers in the United States and Great Britain had built computing machines that could solve large and difficult mathematical problems, graduating from giant mechanical models to much faster and more adaptable electronic models. As the first commercial computers began to appear, and as researchers continued to make computers even faster and more powerful, a fundamental problem remained: memory.

Most people's brains require them to write down each phase of a calculation. Computers could be fed information through rewiring logic circuits or by presenting data and instructions on punch cards or paper tape, with the necessary information encoded through a specific pattern of holes. Like a human trying to solve a complicated math problem, however, computers also needed some kind of "scratch paper" to hold on to important information, such as the different values used in the process of making a calculation.

In the early years of machine computing, there were a few existing means of creating memory. One was a large spinning magnetic drum, similar in concept to the disk memory that is still part of modern computers. Much too slow to take advantage of the lightning-fast pace of electronic calculations, however, the magnetic drum caused computers to slow down. Other methods used cathode ray tubes and mercury-filled tubes. All these options were costly, bulky, and low in capacity.

In the late 1940s, MIT initiated a study called Project Whirlwind to build a real-time

The performance needs of the Whirlwind computer, the first computer ever developed to work in "real time," demanded more and faster memory, which led Jay W. Forrester, Director of the Digital Computer Laboratory at MIT, to develop revolutionary magnetic core memory. Forrester is pictured here during production of "The Search," a CBS special presentation.

computer that could simulate the pace and complexity of airplane flight. For this project, the problem of memory was particularly acute. But researcher Jay Forrester, who headed the project, worked with a team of engineers to create an array of wires intersecting with tiny, magnetic ferrite donuts, or cores. Changing the flow of current to the wires could change the magnetic state of the cores, allowing them to represent 1s or 0s, which in turn allowed the information to be read by the computer.

Enlarged close-up of a core plane (the distance between the rings is approximately one millimeter), showing the matrix of wires used to alter the magnetic polarity of each as a way to store single "bits" of information.

The idea was powerful from several angles. First, the relative simplicity of the system meant costs were lower, and capacity could be increased dramatically compared to other existing data storage methods. Second, information stored in core memory could be accessed quickly. Thanks to the array structure and the multiple wires per core design, information could be written to or retrieved from any location within the structure without having to scroll through data sequentially. This ability to access data randomly from random access memory (RAM) is today implemented through chip technology but remains a component of every computer and electronic device.

Forrester grew up on a Nebraska farm and was a born tinkerer as well as a gifted engineer, manager, and theorist. Members of his MIT team later formed Digital Equipment Corporation (DEC), which mass-produced some of the first minicomputers, giving individual researchers and small companies their own personal computers. At one time DEC ranked second only to IBM in the US computer industry. Inspired by what he learned as a computer architect, Forrester later moved on to a long career as a management visionary, focusing particularly on systems theory.

After filing his patent, Forrester had to deal with lawsuits of other researchers with similar ideas. He and MIT eventually prevailed and, in 1972, just as magnetic core memory was finally being eclipsed by semiconductor memory, the technology that has carried computing into the twenty-first century, Forrester was awarded the IEEE Medal of Honor by the Institute of Electrical and Electronics Engineers for his crucial accomplishment.

Today, core is a distant memory. Early installations of the technology, often measuring only a few hundred bytes, were considered large. This breakthrough technology boosted computing into the space age and nearly into the era of the personal computer.

1969

One Giant Step

MAKING THE MOON LANDING POSSIBLE

On May 25, 1961, President John F. Kennedy called for the United States to land a human on the moon by the end of the decade. The task was so formidable that many questioned the ability of the National Aeronautics and Space Administration (NASA) and US industry to achieve the goal. Not only did the country lack rockets of sufficient size to make the trip and any of the myriad systems needed to sustain humans on such a venture, no one even knew exactly how to navigate a craft from Earth to the moon and then from the moon back to Earth. Finding the right navigation system would be one of the biggest challenges to fulfilling Kennedy's challenge.

Traditional, manual navigational methods used on earth were too slow for a spacecraft traveling thousands of miles an hour and were not accurate enough to prevent an error, which could lead to a collision with the moon or an inadvertent trip into deep space! Guid-

The first steps on the moon were made possible by a huge science and engineering effort that extended the state of the art in many fields, including computing. Neil Armstrong, mission commander, took this photograph of Buzz Aldrin near the leg of the lunar module.

ance from Earth, using powerful computers and teams of astronomers and navigation experts, was a possibility, but the vast distances would introduce communication delays that might be equally dangerous. A third option, providing an onboard computer to manage the transit and return, was appealing, but its success seemed dubious given that most computers of the day filled whole rooms, and even the smallest were typically about the size of a refrigerator: much too large for the capsule intended to transport the first humans.

One organization, the MIT Instrumentation Laboratory (now known as the Charles Stark Draper Laboratory), had achieved miracles with its inertial navigation systems built around electromechanical gyroscope technologies. For example, this system was able to guide an aircraft from takeoff on the East Coast to landing on the West Coast without referencing any external information. The lab had also been experimenting with crafting computers that were simple enough and small enough to assist in guiding missiles to a target. This feat was impressive, but it was quite a simple challenge compared with a round trip to the moon.

The lab took on the challenge of creating the Apollo Guidance Computer (AGC), which became the first computer anywhere to use integrated circuit chips and to rely on two forms of core memory. Built by Raytheon for NASA using MIT's designs, the AGC had computing power comparable to the earliest personal computers, which would not appear on the market until a decade later. It was certainly the most powerful computer of its size ever made. By contemporary standards, it was small, occupying about one cubic foot and weighing seventy pounds; the display and keyboard units (known as the display keyboard [DSKY]) added a few more pounds.

While not as user-friendly as a modern personal computer, the AGC was powerful and adaptable, particularly when equipped with software designed with the algorithms and

Margaret Hamilton, a pioneering programmer at MIT who has been credited with inventing the term "software engineering," was crucial in creating the software for the small computer carried along on the Apollo mission to the moon. The stack of computer printout represents the voluminous documentation that often had to be meticulously reviewed to look for syntax and logic errors in the programs.

calculations needed to navigate to and from the moon and to control rocket thrust. An identical AGC was also included in the Lunar Excursion Module (LEM) to help navigate it from the time it separated from the main capsule in orbit above the moon to its landing on the moon, and then through its relaunch and rendezvous back in orbit.

Leading the software team, which was also responsible for developing the system software needed to help the computer manage itself, was a pioneer programmer, Margaret Hamilton. Already a veteran of other critical programming projects, Hamilton's responsibility, and important success, was building in error detection and recovery capabilities so that, when an inevitable glitch cropped up in software or hardware, the AGC, rather than simply stopping, would be able to regain its direction.

Hamilton's attention to detail and her success in creating a fault-resistant computer, especially within the design constraints imposed by the hardware, would win her accolades, but the crowning glory was how her team's system worked when it mattered most. On the Apollo 11 mission, moments before the LEM was to touch down on the lunar surface, hardware issues in the radar caused a series of overloads to the computer that could have shut it down and aborted the landing. Instead, thanks to Hamilton's system, the computer itself identified the problem it was having and asked for human intervention, thus allowing the landing to continue successfully. The computer and its software would also prove their worth in helping to return astronauts to Earth on the nearly disastrous Apollo 13 mission.

The success of the AGC showcased the power and reliability of emerging chip technology and helped to boost its commercial adoption. Hamilton's work on enhancing the reliability of computers and hardware also proved to be seminal. She contributed designs now used to sustain and manage aircraft and even submarines. Later, she launched two influential software companies, Higher Order Software and Hamilton Technologies, and she is credited as the first person to describe the process of software development as software engineering. Recently she was awarded the Presidential Medal of Freedom for her accomplishments.

2002

Making Robots Part of Our Lives

In popular culture, robots have often been portrayed as cold, potentially destructive analogs of human beings, displaying our worst attributes. But not always. The 1977 film *Star Wars* turned the stereotype on its head by introducing two charming automatons, C-3PO and R2-D2. These likeable and useful robots were to have an impact far beyond the realm of fiction when they inspired young Helen Greiner to make robots real. That motivation led her to MIT, where she earned her undergraduate and master's degrees. She and two others from MIT, Rodney Brooks and Colin Angle, launched iRobot, a robotics company headquartered in Bedford, Massachusetts, northwest of Boston.

Although the company would eventually produce some very sophisticated robotic solutions, it aimed first for simplicity in concept and operation. For example, Genghis, the company's first robot, was basically an electromechanical bug with six legs and a kind of instinctive behavior. Rather than being programmed to perform specific tasks, it was given the ability to perform simple tasks like walking, climbing over obstacles, and following people. Some of its characteristics would later appear in robotic vehicles used to explore Mars.

The company next tackled a more serious task: locating dangerous land mines for the US military. It developed the Ariel robot, which in turn led to a contract with the Defense Advanced Research Projects Agency (DARPA) in 1998 to develop what became PackBot, an adaptable, multipurpose robot designed to scout dangerous locations or even to help disarm bombs. PackBot became available just in time to help comb the wreckage of the World Trade Center. Thousands were eventually deployed with troops in Afghani-

Helen Greiner pursued her dream of creating robots and has had a distinguished career at iRobot and other organizations.

stan and Iraq to explore dangerous locations or to detect or disarm improvised explosive devices (IEDs). PackBots also helped in the Fukushima nuclear disaster, entering locations that humans could not reach or that were too dangerous. For many people throughout the world, it was the first time they saw images of the robots.

But Greiner's vision of a friendly and useful robot was about to take a new smaller and

pool cleaning, and gutter cleaning. In total, the company has sold more than six million home robots.

Today, iRobot, which spun off its military products into a separate division in 2016, continues to be the global leader in consumer-oriented robotic products, employing some nine hundred people. Colin Angle remains as chair and CEO; Rodney Brooks went on to cofound Rethink Robotics, a company based

Genghis (1991) was the first robot developed by iRobot. Intended as a means of testing robotic concepts, it is now part of the collection of the Smithsonian Air and Space Museum.

more affordable form. In 2002, the Roomba robotic vacuum cleaner went on the market. It was not the first robotic vacuum cleaner, but its low cost ($199) and clever method of thoroughly cleaning a room and returning to its docking station to recharge endeared it to consumers. The company continued to improve the design, creating several companion products to perform tasks like floor mopping,

in Boston that aimed to create low-cost robots to collaborate with humans on work tasks. Helen Greiner left the company in 2008 and has since become an industry leader. Among other honors, she received the 2008 Anita Borg Institute Women of Vision Award. Today, Boston's multifirm robotics industry is one of the leading clusters globally in this industry.

2011

"Siri, Where Did You Come From?"

SPEECH RECOGNITION, JAMES AND JANET BAKER, DRAGON SYSTEMS, AND NUANCE

Machines that listen, interpret, and respond were considered science fiction just a short time ago. But science fiction became science fact just a decade into the twenty-first century when the virtual assistant Siri was integrated into products from Apple Inc. and it and others became part of the lives of millions of people.

Siri is made up of two fundamental technologies. Its thinking was developed originally by Stanford Research Institute's (SRI's) Artificial Intelligence Center (thus the name Siri). Its ability to recognize and interpret human speech was developed originally by Dragon Systems, of Newton, Massachusetts, whose assets and technology were later acquired by the Massachusetts-based company Nuance Communications. With its uncanny ability to recognize words correctly, and its further ability to interpret those words and respond knowledgeably, Siri has become integral to Apple and its customers. Asking for help from a machine—and getting it—now seem completely natural and unremarkable to millions, even billions, of individuals.

Machines that can interpret and even understand human speech are now common, but their existence rests on many years of innovation. The most significant leap forward was made possible by a husband and wife team who focused on the problem for decades, gradually harnessing algorithms to decipher the complexities of human speech. By the mid-twentieth century, researchers had begun to realize that modern computers could be the key to making speech-to-text and even machine understanding a reality. But how? Most researchers thought machines needed to determine speech components (acoustics, syntax, semantics, etc.) separately before they could interpret them into text; however, this approach ultimately proved unsatisfactory. The major breakthrough came about by using stochastic mathematical algorithms to model speech and language components jointly to convert speech into text.

In the 1970s, graduate students James Baker, a mathematician, and Janet M. Baker, a biophysicist, were working on novel speech recognition concepts at Rockefeller University in New York City but accepted an invitation to relocate to Carnegie-Mellon in Pittsburgh, which had more abundant computing resources. There, Jim mapped out the methodology and implemented a probabilistic speech recognition system they dubbed DRAGON. HARPY, a refined version of DRAGON, took first place in the five-year multi-institutional Defense Advanced Research Projects Agency (DARPA) Speech Understanding Research program in 1976.

After several years working in speech research for IBM and Verbex, Jim and Janet founded Dragon Systems in 1982 to advance

Jim and Janet Baker began their collaboration to make possible machine recognition of human speech in the 1970s.

the technology and bring state-of-the art speech recognition to users on affordable devices. Without any venture capital or external financing (until twelve years later), they bootstrapped the company with thirty thousand dollars of their own savings as well as sweat equity. Focused on their mathematical methodology and computational efficiency, their company in Newton, Massachusetts (a few miles from Boston), eventually took off. It offered a series of commercial products that could run on the personal computers of the time. As the couple later said, "We wanted to do something that would be practical and useful and more than a paper on a library shelf."

From its introduction, Dragon was a hit, even though it had limited capabilities on the computers of the time, and early versions required some initial interaction with each speaker to make it work optimally. With each generation of more powerful computers and with each successive software upgrade, however, Dragon became more capable and, for many users, indispensable. In 1997, their company, Dragon Systems, launched Dragon Naturally Speaking, the first general-purpose, continuous dictation system, which was showered internationally with awards and accolades and performed well in the marketplace.

Janet and Jim's road to success took a detour, however, when they agreed to be acquired for about $600 million by a European company, the high-flying Lernout & Hauspie Speech Products (L&H), which was acquiring companies across the speech field. However, the widely respected company turned out to have been built on widespread fraudulent practices, much like Enron in the United States. As a result, when L&H went bankrupt in 2000, the Bakers, who had reluctantly accepted an all stock purchase, ended up with nothing—not even the rights to their own technology.

Eventually, as Janet Baker has explained, a small portion of their company's original value was recouped through lawsuits. Dragon Systems technologies, patents, and products, included in a Lernout & Hauspie bankruptcy auction, were ultimately acquired by Scansoft, a Burlington, Massachusetts, company focused on character recognition. Later renamed Nuance Communications, the

Jim Baker using speech recognition software on an early IBM PC. The same technology has been steadily improved ever since and is now embedded in services such as Siri.

company has since developed and globally licensed its Dragon-based technologies and products across many fields (medical, legal, consumer, etc.) including licensing it for use in Siri.

In 2012, at which point their technology had become ubiquitous in Siri, the Bakers were jointly awarded the Institute of Electrical and Electronics Engineers (IEEE) James L. Flanagan Speech and Audio Processing Award for "fundamental contributions to the theory and practice of automatic speech recognition."

But perhaps just as impressive is that Siri has become a household word, and we take no notice of people talking *to* their phones, devices, and microphones most anywhere, to command them and ask questions. Clearly, Janet and James succeeded beyond their own wildest dreams.

FINISH WITH A SMILE

1964 Emoticon Number 1
THE FIRST SMILEY

For a language to thrive, it requires a common vocabulary, and every human language has one. Languages for computers also have their lexicons, as does the language of emoticons. Emoticons are pictographs composed from a linguistic character set, and emojis are pictographs that illustrate specific emotions or ideas. Both have become an international language for the twenty-first century, conveying quickly and concisely a range of human emotions and concepts. Emoticons and emojis have created an ability to give an electronic message, a medium sometimes seen as sterile and emotionless, an emotional dimension that text and email text cannot.

With the rise of short message service (SMS) texting, the importance of emoticons and emojis has grown immensely, but how did this new universal language get its start? It would be fruitless to suggest any single point of origin. In the media-rich decade of the 1960s, in which images (think Beatles haircuts, the peace symbol, and the rainbow palette of the youth movement) grew to be universally recognizable almost overnight, one image began as a corporate symbol and has had particular staying power. That symbol, the smiley face, is still with us, in everything from folk art to advertising. It is nearly indistinguishable from the most basic emoji that emerged in the late 1990s.

In the early 1960s, the State Mutual Life Assurance Company, based in Worcester, Massachusetts, had just acquired another insurance company based in Ohio and had a familiar problem after the merger: sagging employee morale. Individuals were trying to understand their role in the new and larger organization and were perhaps wondering about their futures. Harvey Ball, a graphic designer with his own small advertising business, was hired to provide a solution: a symbolic shorthand for encapsulating the more upbeat tone that management at the insurance company wished to convey to its employees.

Harvey Ball's creation was an unambiguous smiley face rendered as simply as possible, with a circle, two elongated dots for eyes, and a curving smile—all made strikingly cheery through its use of black ink on a bright yellow background. Executives loved the design and made it the centerpiece of internal communication with thousands of employees. It later migrated to communication with customers as well, reaching millions of people in a short time. Neither Ball nor State Mutual had taken the time to trademark the design, so in short order imitators were popping up all over, in print, on buttons, and on other three-dimensional objects. By the early 1970s, at least fifty million buttons had been produced, making the smiley face an international icon. A French firm, which trademarked a similar design,

Harvey Ball's simple design for a smiling face, designed to boost corporate morale at an insurance company, had already become a global phenomenon by the early 1970s, when this photo was taken.

went on to sell millions of products also emblazoned with the stylized happy face.

Users of electronic devices and computers gravitated toward shorthand messaging with clever combinations of punctuation marks, and soon they were adding emojis, including as the most basic emoji character, a face strongly reminiscent of the widely known Harvey Ball/State Mutual smiley face. Today, billions of emojis are transmitted daily—happy, sad, funny, and more—each a descendant of the crystal clear, colorful, and instantly recognizable image from a clever designer.

For his design, Ball received about forty-five dollars (just over three hundred in current monetary value). Ball, who died in 2001, reportedly had "no regrets" for not somehow earning more money from his innovation, though much later, in 1999, he established the World Smile Corporation to support World Smile Day, a nonprofit charitable trust that supports children's causes. In recognition of the impact of his perpetually cheery image, his hometown of Worcester established the Smiley Face Trail, linking Broad Meadow Brook Sanctuary and Blackstone River Bikeway through the Harvey Ball Conservation Area.

Conclusion

As the introduction to this book shows, there are consistent, unifying reasons why the human activity of innovation has been so successful in the Boston metropolitan area. The key innovation drivers, especially the "bump and connect" process, offer a path for those determined to muster a countervailing energy that can change an industry or a society—or to tease from nature the hidden secrets of a merciless disease.

Those key drivers—prevalence of local finance, forceful entrepreneurs, the interconnectedness of innovators who listened and learned from each other, and local and global demand—are behind Boston's innovation success. But equally important has been Boston's ability to bounce back from decline. Four times over the last four centuries, the Greater Boston economy collapsed. Yet each time, the region's culture and economy responded to collapse with a new wave of innovation that pushed it back into the highest levels of innovative creativity, including the most recent wave, powered by technology, biotech, robotics, and finance.

Many cities have one peak of innovation during a few hundred years, perhaps two, but few come back from the brink four times. It is not easy. Yet what the Boston region has done can be done by other cities, other regions, other nations. Innovation's drivers play differently at different times and against different challenges. The interplay of drivers will be unique to every geography. But as *Boston Made* demonstrates, innovation has a massive power to make life and society better. It is something our complex world needs and something well worth pursuing.

This book has been a kind of dare, a gamble that readers will engage with a complex but fascinating subject if provided with many different entry points. Its stories are intriguing, entertaining, and in some cases very moving. They range from the fight to abolish slavery to the redefinition of marriage, from the effort to bring the ancient craft of sailing to its highest perfection to the work of scientists to alter molecular structures as they develop powerful cures to humankind's worst diseases.

Boston's innovations range across time, space, and social structures. Together they create a big picture while allowing us to appreciate the private dramas and challenges of individuals determined to make a difference, whether by inventing rockets to reach the moon or convincing a frightened city to take an informed risk by embracing the words of an enslaved man as a source of salvation at a time of plague.

So, it is hoped that these Boston stories—based on over twenty years of research and selected from a superset of nearly five hundred significant breakthroughs—offer not only fascinating snapshots of history, but also a collective sense of how it is possible for a city to reinvent over and over again, and how individuals, supported and nurtured by a creative environment, can overcome technical and legal barriers, as well as the doubt and hostility of naysayers, to create and invent and, in doing so, change the world.

Bibliography

Adams, Russell. *Boston Money Tree: How Proper Men of Boston Made, Invested, and Preserved Their Wealth from Colonial Days to the Space Age.* New York: Crowell Company, 1977.

Adams, Russell. *King Gillette: The Man and his Wonderful Shaving Device.* Little Brown, Boston 1978.

Allen, David Grayson. *Investment Management In Boston: A History.* Amherst: University of Massachusetts Press, 2015.

Allison, Robert. *A Short History of Boston.* Carlisle, Massachusetts: Commonwealth Editions, 2004.

Bissell, Don. *The First Conglomerate: 145 Years of Singer Sewing Machine Company.* Brunswick, Maine: Audenreed Press, 1999.

Blatt, Martin H., Thomas J. Brown, and Donald Yacovone, eds. *Hope & Glory: Essays on the Legacy of the 54th Massachusetts Regiment.* Amherst: University of Massachusetts Press, 2001.

Brooks, John. *Telephone: The First Hundred Years.* New York: Harper Brothers, 1976.

Bruce, Roger R., ed. *Seeing the Unseen: Dr. Harold E. Edgerton and the Wonders of Strobe Alley.* Cambridge: MIT Press, 1994.

Cohen, I. Bernard. *Howard Aiken: Portrait of a Computer Pioneer.* Cambridge: MIT Press, 2000.

Cooke, Gilmore. *The Existential Joys of Fred Stark Pearson, Engineer, Entrepreneur, Envisioner.* New Providence, New Jersey: Bowker, 2020.

Cooke, Robert. *Dr. Folkman's War: Angiogenesis and the Struggle to Defeat Cancer.* New York: Random House, 2001.

Dowling, Tim. *Inventor of the Disposable Culture: King Camp Gillette 1855–1932.* London: Short Books, 2002.

Earls, Alan, *Route 128 and the Birth of the Age of High Tech*. Mount Pleasant, South Carolina: Arcadia Publishing, 2002.

Earls, Alan, and Robert E. Edwards. *Raytheon Company: The First Sixty Years*. Mount Pleasant, South Carolina: Arcadia Publishing, 2005.

Forbes, Esther. *Paul Revere and the World He Lived In*. New York: Houghton Mifflin Harcourt, 2018.

Fouché, Rayvon. *Black Inventors in the Age of Segregation: Granville T. Woods, Lewis H. Latimer, and Shelby J. Davidson*. Baltimore: The Johns Hopkins University Press, 2003.

Glaeser, Edward L. "Reinventing Boston: 1630–2003." *Journal of Economic Geography* 5 (2) (2005): pages 119–153.

Goddard, Stephan B. *Colonel Albert Pope and His American Dream Machines*. Jefferson, North Carolina: McFarland Publishing, 2000.

Gozemba, Patricia, and Karen Kahn. *Courting Equality: A Documentary History of America's First Legal Same-Sex Marriages*. Boston: Beacon Press, 2007.

Grosvenor, Edwin S., and Morgan Wesson. *Alexander Graham Bell: The Life and Times of the Man Who Invented the Telephone*. New York: Harry N. Abrams, 1997.

Gupta, Udayan, ed. *The First Venture Capitalist: Georges Doriot on Leadership, Capital, and Business Organization*. Calgary, Canada: Gondolier, 2004.

Gura, Philip F. *American Transcendentalism: A History*. New York: Hill and Wang, 2008.

Hafner, Katie, et al. *Where Wizards Stay Up Late: The Origins of the Internet*. New York: Simon and Schuster, 1999.

Isaacson, Walter. *Benjamin Franklin, An American Life*. New York: Simon & Schuster, 2003.

Johnson, Steven. *Where Good Ideas Come From: The Natural History of Innovation*. New York: Riverhead Books, 2010.

Kolodony, Kelly Ann. *Normalites: The First Professionally Prepared Teachers in the United States*. Charlotte, North Carolina: Information Age Publishing, 2014.

Krensky, Stephen. *What's the Big Idea? Four Centuries of Innovation in Boston*. Watertown, Massachusetts: Charlesbridge, 2008.

Mabee, Carleton. *The American Leonardo: A Life of Samuel F. Morse*. New York: Purple Mountain Press, 2000.

Maier, Pauline, et al. *Inventing America: A History of the United States*. New York: Norton, 2006.

Marshall, Megan. *Margaret Fuller: A New American Life*. New York: Mariner Books, 2013.

Marx, Christy. *Grace Hopper: The First Woman to Program the First Computer in the United States*. New York: Rosen Publishing Group, 2003.

McElheny, Victor K. *Insisting on the Impossible: The Life of Edwin Land*. Cambridge, Massachusetts: Perseus Books, 1998.

McKay, Richard C. *Donald McKay and His Famous Sailing Ships*. Toronto: Dover Press, 1995.

McKibben, Gordon. *The Cutting Edge: Gillette's Journey to Global Leadership*. Boston: Harvard Business School Press, 1998.

Morgenroth, Lynda. *Boston Firsts: 40 Feats of Innovation and Invention That Happened First in Boston and Helped Make America Great*. Boston: Beacon Press, 2006.

Most, Doug. *The Race Underground: Boston, New York, and the Incredible Rivalry that Built America's First Subway*. New York: St. Martin's Press, 2014.

Bibliography

Murphy, Robert. *An Introduction to AI Robotics*. Cambridge: MIT Press, 2007.

Norman, Winifred Latimer, and Lily Patterson. *Lewis Latimer (Black Americans of Achievement)*. Broomall, Pennsylvania: Chelsea House Publishers, 1993.

Peterson, Mark. *The City-State of Boston: The Rise and Fall of an Atlantic Power, 1630–1865*. Princeton, New Jersey: Princeton University Press, 2019.

Pridmore, Jay, and Jim Hurd. *The American Bicycle*. Osceola, Wisconsin: Motorbooks International, 1995.

Seaburg, Carl. *The Ice King: Frederic Tudor and His Circle*. Mystic, Connecticut: Mystic Seaport Museum, 2003.

Seasholes, Nancy. *The Atlas of Boston History*. Chicago: University of Chicago Press, 2019.

Struik, Dirk J. *Yankee Science in the Making*. New York: Collier Books, 1968.

Vrabel, Jim. *When in Boston: A Time Line & Almanac*. Boston: Northeastern University Press, 2004.

Walshok, Mary Lindenstein, and Abraham Schragge. *Invention and Reinvention: The Evolution of San Diego's Innovation Economy (Innovation and Technology in the World Economy)*. Palo Alto: Stanford University Press, 2014.

Wright, Conrad, and Katheryn P. Viens, eds. *Entrepreneurs: The Boston Business Community 1700–1850*. Boston: Massachusetts Historical Society, 1997.

Illustration and Photo Credits

Page 13: *Bunker Hill Monument/Zakim Bridge*, Mark Garfinkel, used with permission
Page 20: *Sawyer robot*, Mass Robotics © 2019, used with permission
Page 25: *Kendall Square, Cambridge*, aerial by lesvants.com, used with permission
Page 28: *Legal notes by William Cushing about the Quock Walker case*, Collection of the Massachusetts Historical Society
Page 28: *portrait of Elizabeth Freeman ("Mumbet")*, Collection of the Massachusetts Historical Society
Page 29: *Walker's Appeal*, Collection of the Massachusetts Historical Society
Page 30: *Liberator masthead*, public domain
Page 30: *brutally scarred slave*, public domain
Page 31: *Henry David Thoreau*, public domain
Page 32: *Margaret Fuller*, Getty/Archive Photos
Page 32: *Walden Pond*, public domain
Page 33: *54th Regiment poster*, Collection of the Massachusetts Historical Society
Page 33: *Private William J. Netson*, Collection of the Massachusetts Historical Society
Page 34: *charge by the 54th Regiment*, public domain
Page 35: *Margaret Marshall and Mary Bonauto*, Susan Symonds, Infinity Portrait Design/Mainframe Photographics, used with permission
Page 36: *marriage at City Hall, Cambridge, Massachusetts*, Marilyn Humphries, used with permission
Page 38: *sledding on Boston Common*, public domain
Page 39: *Parkman Bandstand on Boston Common*, Elizabeth Jordan, Friends of the Public Garden, used with permission

Illustration and Photo Credits

Page 40: *old image of Boston Latin School*, Boston Latin School Association, used with permission

Page 41: *Boston Latin School*, Boston Latin School Association, used with permission

Page 43: *Harvard College in the eighteenth century*, Collection of the Massachusetts Historical Society

Page 44: *Harvard Science Center*, Alan R. Earls, used with permission

Page 46: *statue of Lexington farmer*, public domain

Page 46: *Comitia Americana medal, Washington before Boston, 1776*, Collection of the Massachusetts Historical Society

Page 48: *John Adams*, Collection of the Massachusetts Historical Society

Page 49: *Massachusetts constitution*, Collection of the Massachusetts Historical Society

Page 51: *Framingham Normal School students*: Framingham State University Archives, used with permission

Page 53: *old map of Boston*, public domain

Page 53: *smallpox poster*, public domain

Page 54: *painting commemorating the first use of ether at Massachusetts General Hospital*, MGH Archive, used with permission

Page 55: *Massachusetts General Hospital in the nineteenth century*, public domain

Page 57: *Dr. Sidney Farber*, AP Photo

Page 59: *The First Successful Organ Transplantation in Man, 1996 oil on linen, The Countway Library of Medicine Harvard Medical School, Boston*, Joel Babb, used with permission

Page 61: *Raymond Kurzweil with his reading machine*, Getty/Bettmann

Page 63: *Biogen sign*, Alan R. Earls, used with permission

Page 65: *Dr. Judah Folkman*, Marsha Moses, PhD, used with permission

Page 66: *The "Ice King," Frederic Tudor*, the Bostonian Society, used with permission

Page 67: *ice-cutting in the nineteenth century*, public domain

Page 69: *painting of Charles Goodyear discovering vulcanization*, courtesy of Goodyear Tire & Rubber Company

Page 71: *Elias Howe*, public domain

Page 72: *sewing machine patent drawing*, public domain

Page 73: *Gillette ad*, public domain

Page 74: *baby Gillette ad*, public domain

Page 75: *Toll House, Whitman, Massachusetts*, public domain

Page 76: *chocolate chip cookies*, Getty/LightRocket

Page 78: *frozen food aisle*, Getty/LightRocket

Page 80: *chef with Radarange*, courtesy of Raytheon Company

Page 81: *Katharine Dexter McCormick*, public domain

Page 81: *Margaret Sanger*, public domain

Page 82: *Dr. Min Chueh Chang*, Science Source, used with permission

Page 82: *Dr. Gregory Pincus*, AP Photo

Page 83: *woman holding birth control pills*, Getty/Bettmann

Page 84: *Columbia and Washington medal*, Collection of the Massachusetts Historical Society

Page 85: *voyage of the Columbia*, public domain

Page 87: *ARD annual report*, Baker Library, Harvard, used with permission
Page 89: *image of Fidelity Daily Income Trust*, courtesy of Fidelity Investments © FMR LLC, all rights reserved
Page 90: *ad for Fidelity Daily Income Trust*, courtesy of Fidelity Investments © FMR LLC, all rights reserved
Page 92: *first steam excavator*, public domain
Page 93: *1898 steam shovel*, public domain
Page 94: *whale-hunting painting*, public domain
Page 95: *statue of Lewis Temple*, Ethan McCarthy Earls, used with permission
Page 96: *Donald McKay*, public domain
Page 97: *painting of Flying Cloud clipper ship*, MHP Enterprises, used with permission
Page 99: *construction of the Boston underground*, public domain
Page 101: *Boston trolley*, public domain
Page 102: *James Naismith*, Getty/Bettmann
Page 103: *Larry Bird in action*, Lipofskyphoto.com, used with permission
Page 104: *statue of Cy Young*, courtesy of Northeastern University
Page 105: *World Series game at Huntingdon Avenue field, Boston*, public domain
Page 106: *movie projector*, public domain
Page 107: *The Return of Frank James*, Getty/Moviepix
Page 109: *image of electronic paper*, courtesy of E Ink Corporation
Page 111: *First American newspaper, Publick Occurrences*, Collection of the Massachusetts Historical Society
Page 112: *actor portraying Bell with centennial phone*, courtesy of AT&T Archives and History Center
Page 113: *early switchboard operators*, public domain
Page 115: *ARPANET map*, Kira Beaudoin, used with permission
Page 116: *Bob Frankston and Dan Bricklin flanking an IBM PC*, © www.jimraycroft.com 1982, used with permission
Page 117: *VisiCalc screenshot*, public domain
Page 119: *Tripadvisor sign*, Getty/Archive Photos
Page 120: *Mark Zuckerberg*, Getty/SOP Images
Page 122: *Robert Goddard and his rocket*, courtesy of Clark University Archives
Page 123: *Marshall Space Flight Center rocket*, courtesy of National Aeronautics and Space Administration
Page 125: *Jay Forrester and Whirlwind computer*, Getty/CBS Photo Archive
Page 126: *core memory innards*, public domain
Page 127: *Buzz Aldrin on the moon*, public domain
Page 128: *Margaret Hamilton*, courtesy of Draper Labs
Page 130: *Helen Greiner and robot*, courtesy of iRobot
Page 131: *Genghis robot*, courtesy of iRobot
Page 133: *Janet and James Baker*, courtesy of Janet M. Baker
Page 134: *James Baker*, courtesy of Janet M. Baker
Page 136: *Harvey Ball and his "smiley,"* courtesy of Worcester Historical Museum

Index

54th Massachusetts Infantry
   Regiment, 33–34

A

@, first use in email, 115
abolitionists, 33
ACLU, *See* American Civil
   Liberties Union
ACS, *See* American
   Colonization Society
Adams, John, 27, 48
Adams, Samuel, 19, 41, 48–49
Advanced Research Projects
   Agency (ARPA), 114
Advanced Research Projects
   Agency Network, *See*
   ARPANET
African American, 3, 8, 16, 27,
   29, 33–34, 51
   enslaved, 16
   free-born, 29
African American regiment,
   See 54th Regiment
African American freedom, 33
AGC, *See* Apollo Guidance
   Computer
Akamai Technologies, 6
Albert, J.D., 108
Amazon Kindle, 108
American Civil Liberties Union
   (ACLU), 36
American Colonization Society
   (ACS), 29
American League, 104–5
American Research and
   Development, 87
American Revolution, 2, 27,
   33, 39, 46, 48, 84–85, 97
Amherst College, 77
Andrew, John A., 33
angiogenesis, 64–65
Angle, Colin, 130–31
Anita Borg Institute
   Women of Vision
   Award, 131
Antioch College, 51
Apollo Guidance Computer
   (AGC), 128–29
ARD, *See* America Research
   and Development
Ariel robot, 130
Armstrong, Neil, 127
ARPANET (Advanced
   Research Projects Agency
   Network), 114–15
Ashley, John, 27
Auburn, Massachusetts, 123
Avastin, 65
Avonex, 62
Azorean Portuguese, 95

B

Baird, Bill, 81, 83
Baker, Janet, 133
Baker, Jim, 134
Ball, Harvey, 135–36
Baltimore, 85, 98
bank, first state chartered, 84–85
Barnes & Noble's Nook, 108
basketball, invention of, 102–3
Battery Ventures, 121
Battle of Bunker Hill, 13
BBN, *See* Bolt, Beranek and Newman
Bell, Alexander Graham, 112–13
Bell Labs, 60
Bell telephone system, 113
Biogen, 23–24, 62–63
biotechnology ('biotech'), 5–7, 15, 23–26, 61–63, 86, 137
Bird, Larry, 103
Birds Eye Frozen Foods Company, 77
Birdseye, Clarence, 77–78
birth control pill, 3, 81–83
Bodenhorn, Howard, 85
Bolt, Beranek and Newman (BBN), 114–15
Bonauto, Mary, 35, 37
Boston Braves, 56
Boston Celtics, 103
Boston Common, 38–39, 45, 99
Boston Computer Society, 11, 117
Boston Garden, 103
Boston Globe, 111
Boston Herald, 111
Boston History & Innovation Collaborative, 6
Boston Latin School, 9, 32, 40–41
Boston News-Letter, 110
Boston Tea Party, 18
Boston University, 4, 83, 112
Boston's HarborWalk, 67
Bowdoin, James, 48–49
Boylston, Zabdiel, 52–53
Bricklin, Dan, 116
Brooke, Peter, 26
bump and connect, 12–15, 23
bump and connect process, 137
Bunche, Ralph, 3
Bunker Hill, 13, 46
Burton, Virginia Lee, 92

C

Cambridge, Massachusetts, 5, 15, 23–24, 31, 36, 44, 46, 52, 62, 67, 71–72, 110, 115
cancer, 56–57, 64–65, 87
Canton, Massachusetts, 91
Carnegie-Mellon, 132
Challenger, 98
Chang, Min Chueh, 81–82
Chapman, Oliver, 92
Charlestown, Massachusetts, 67
chemotherapy, 56–57
Children's Cancer Research Foundation, 56
Children's Hospital, 56
China, 16, 81
  trade with, 19, 84–85
chocolate chip cookie, 2, 75–76
Civil War, 19, 30, 33–34, 51, 98, 107, 112
Clark University, 122, 124
Clinton, Massachusetts, 93
clipper ships, 5, 10, 97–98
cod, 11
color movies, 106–7
Columbia (ship), 84–85
Comiskey, Barrett, 108
Comstock, Daniel, 106
Concord, Massachusetts, 18, 31–32, 45–46
Continental Congress, 46–47, 49
core memory, 126, 128
Cornell University, 103
Cotton, John, 40–41
Cressy, Josiah Perkins, 98
Crick, Francis, 14
Cushing, Richard Cardinal, 58

D

Dana, Charles A., 57
Dana-Farber Cancer Institute, 57
DARPA, *See* Defense Advanced Research Projects Agency
Dartmouth College, 103
Davidson, Olivia, 51
DEC, *See* Digital Equipment Corporation
Defense Advanced Research Projects Agency (DARPA), 130, 132
Digital Equipment Corporation (DEC), 87, 116, 126
Disney, Walt, 107
DNA, 14–15, 62
Dorchester Heights, Massachusetts, 47
Doriot, Georges, 26, 87–88
Douglass, Frederick, 33
Dragon Systems, 132–33
Drivers of innovation
  five, 6–7, 23
  key, 137

E

East Anglia, 40
East Boston, Massachusetts, 10, 96
Edison, Thomas, 99, 113
e-ink, 108
electronic paper, 108
Emerson, Ralph Waldo, 31–32
Ether Dome, 55
excavator, steam-powered, 9, 92

F

Facebook, 120–21
Farber, Sidney, 56–57
Farrakhan, Louis, 41
Fidelity Daily Income Fund, 89
Fidelity Investments, 12, 89–90
Flying Cloud, 96
Folkman, Judah, 8, 64–65
foods, frozen, 22, 77–78
Forrester, Jay W., 125
Fort Sumter, 34
Fort Ticonderoga, 47
Fort Wagner, 34
Framingham State Normal School, 51
Framingham State University, 75
Franklin, Benjamin, 41, 52, 111
Franklin, James, 111
Franklin, Rosalind, 14
Frankston, Bob, 116
Freedom to Marry Coalition of Massachusetts, 36
Freeman, Elizabeth (Mumbet), 16, 27–28
French, Daniel Chester, 46
Fuller, Margaret, 32
Fylstra, Dan, 116

G

Garrison, William Lloyd, 30
Gaucher's disease, 6
Gay and Lesbian Advocates and Defenders (GLAD), 36–37
Genghis robot, 130–31
Gilbert, Walter, 23, 62
Gillette, King, 9, 73
Gillette invention, 74
GLAD, *See* Gay and Lesbian Advocates and Defenders
Globe Locomotive Works, 92
Glorikian, Harry, 26
Goddard, Robert, 122–24
Goddard Space Flight Center, 124
Goodridge v. Dept. of Public Health, 35, 37
Goodyear, Charles, 8, 68–70
Greiner, Helen, 130–31
Guggenheim Foundation, 124

H

Hamilton, Margaret, 128–29
Hamilton Technologies, 129
Hancock, John, 41, 49
harpoon, 94–95
Harris, Benjamin, 110
Harvard, John, 43
Harvard Business School, 11, 87–88, 116
Harvard College, 4, 25, 32, 43–44, 110, 120–21
  founding of, 42–43
Harvard Divinity School, 43
Harvard Medical School, 23, 64, 82
Harvard University, 23, 25, 43, 44, 62,
Harvey Ball Conservation Area, 136

Heart, Frank, 114
Herrick, Richard, 58–59
Herrick, Ron, 58
High Voltage Engineering Company, 87
Howe, Amasa Bemis, 72
Howe, Elias, 71–72
Hubbard, Gardiner Green, 113
Hubbard, Mabel, 113
Human Genome Project, 62
Hutchinson, Anne, 42

I

Ice King, *See* Tudor, Frederic
IMP, *See* Interface Message Processor
innovation, waves of, 2–3, 6, 9, 15, 17, 138
inoculation, 52–53
Interface Message Processor (IMP), 114–15
International Business Machines Corporation (IBM), 116
internet, 114–16
iRobot, 20, 130–31

J

Jacobs, Lawrence, 62
Jacobson, Joseph, 108
Jet Propulsion Laboratory, 124
Jimmy Fund, 56
Jobs, Steve, 96, 117
Johnson, Edward (Ned), 89–90
Johnson, Magic, 103
Johnson, Steven, 14
Joyce, John, 74
JPL, *See* Jet Propulsion Laboratory

K

Kalmus, Herbert, 106
Kaufer, Steve, 118
Kendall Square, 15, 23–26, 44, 67
Kennedy, Edward, 24
Kennedy, Joseph, 41
Killilea, Henry, 104
King, John, 41
King, Martin Luther, Jr., 32
Kirsner, Scott, 4
Knox, Henry, 47
Kurzweil, Ray, 60–61
Kurzweil K250, 61
Kurzweil Reading Machine, 60

L

Lady Washington (ship), 84
Langer, Robert, 64
Ledyard, John, 19
LEM, *See* Lunar Excursion Module
leukemia, 58
Lexis-Nexis, 108
LGBTQ, 35–37
Liberator, 30
Liberia, 29
Liberty Tree, 45
Licklider, J.C.R., 114
Logan International Airport, 3, 23
Los Angeles Lakers, 103
Lotus Software, 117
Lowell, Massachusetts, 71
Lunar Excursion Module (LEM), 129

M

Mandela, Nelson, 32
Mann, Horace, 50–51
Mann, Mary Peabody, 50
marriage, same-sex, 35–36
Marshall, Margaret, 35, 37
Marshall Space Flight Center, 123
Massachusetts Bank, 84
Massachusetts Bay Colony, 17–18, 40, 42
Massachusetts Constitution, 16, 37, 49
 new, 16, 27–28, 48
Massachusetts General Hospital (MGH), 23, 54–55
Massachusetts Institute of Technology (MIT), 4, 7, 9, 23, 25, 44, 60–65, 74, 81–82, 87, 106, 108, 112, 114, 125–26, 128, 130
Massachusetts Life Sciences Act, 23
Massachusetts State Board of Education, 50
Massachusetts State Supreme Judicial Court (SJC), 27, 35, 37
Mather, Cotton, 52
McCormick, Katharine, 81–82
McKay, Donald, 10, 96–97
Melville, Herman, 95
MGH, *See* Massachusetts General Hospital
Milken Institute, 5
Minimum Viable Product (MVP), 50
Moby Dick, 95
Monroe, Lewis, 112
moon flight, 124
Morton, William, 54–55
multiple sclerosis (MS), 2, 15, 62
Mumbet, *See* Freeman, Elizabeth
Murray, Joseph, 3, 58–59

N

Naismith, James, 102–3
National Aeronautics and Space Administration (NASA), 124, 127–28
National Federation of the Blind (NFB), 60
National Institutes of Health (NIH), 25
National League, 104
National Organization on Rare Diseases, 24
Nestlé, 75–76
Netson, William J., 33
New Amsterdam, 40, 42
New Bedford, Massachusetts, 95
New England, 2, 19, 56, 89, 94, 98, 110
New-England Courant, 111
New York City, 40, 68, 98, 101, 113, 132
New York Times, 123
New York Yankees, 105
Newburyport, Massachusetts, 10, 97
Newton, Massachusetts, 31, 79, 132–33
NFB, *See* National Federation of the Blind
Nickerson, William, 9, 74
NIH, *See* National Institutes of Health
Nobel Prize, 3, 5, 14, 23–24, 44, 59, 62
NORD, *See* National Organization on Rare Diseases
Northeastern University, 4, 105
Novartis, 16

Index

O
Onesimus, 52–53
organ transplant, kidney, 58–59
Orphan Drug Act, 24–25
Otis, William, 9, 91–92

P
Pacific Northwest, 84
PackBots, 130–31
Panic of 1837, 69
Panic of 1857, 19
Pearson, Fred, 100
Peirce, Cyrus, 50
Peter Bent Brigham Hospital, 58
Philadelphia, 29, 46–47, 49, 55, 68, 85, 92, 98, 111, 113
Pickering, Edward C., 112
Pill, the, *See* birth control pill
Pincus, Gregory, 81–82
Procter & Gamble, 73
Project Whirlwind, 125
public school, first, 40–41
Publick Occurrences, 110

R
radar, 21, 79, 129
Radarange, 1, 79–80
random access memory (RAM), 126
Raytheon Company, 1, 21, 79, 115, 128
reading machine, 60–61
recolonization, advocated, 30
Redstone, Sumner, 41
Reuff, Martin, 14
Revere, Paul, 45, 91
Revere Copper Company, 91
Revolutionary War, 19

Rock, John, 81–82
rockets, first liquid-fuel, 123–24
Roomba robotic vacuum, 131
Rubin, Jerome, 108

S
safety razor, 73–74
Salem, Massachusetts, 84
Sanger, Margaret, 81
Sanofi Genzyme, 24
Saverin, Eduardo, 121
Sedgewick, John, 27
sewing machines, 72
Shannon, Claude, 114
Shanny, Nick, 118
Sharp, Philip, 23, 62
Shaw, Robert Gould, 33–34
Sheffield, Massachusetts, 27, 31, 33
Sidney Farber Cancer Institute, 56
Silicon Valley, 18, 22, 121
Singer, Isaac, 72
Siri, 132, 134
slavery, 2, 27–28, 31, 33, 49, 53, 138
smallpox, 19, 47, 52–53
smallpox inoculation, 53
smiley, 135–36
South Boston, Massachusetts, 92
South Carolina, 34
Souther, John, 92
Spencer, Percy, 1, 79
Sprague, Frank, 100
Springfield, Massachusetts, 103
Springfield College, 102
Stag Hound (ship), 96
Stamp Act, 45
State Mutual Life Assurance Company, 135

state of Massachusetts, 2, 28, 45, 49–50, 85
Steinert, Nick, 118
Stone, Abraham, 81
subway, 20, 23, 100–1
 first, 99, 101
Supreme Judicial Court, *See* Massachusetts Supreme Judicial Court

T
Technicolor, 106–7
Temple, Lewis, 95
Temple Toggle Iron, 95
Thiel, Peter, 121
Thomson-Houston Company, 100
Thoreau, Henry David, 31–32
Thursday's Child (ship), 97
Toll House, 75–76
Tomlinson, Ray, 115
Train, Enoch, 10, 96
transcendentalism, 31–32
Tripadvisor, 119
Tudor, Frederic, 66
Tudor Wharf, 67

U
University of California, 115
University of Chicago, 103
University of Kansas, 103
University of Massachusetts Boston, 4
USS Constitution, 91

V
Vane, Henry, 41
Vellucci, Alfred, 24
venture capital, 10, 63, 86–88, 133
VisiCalc, 11, 116–17
vulcanized rubber, 68–70

W

Wagner, Honus, 74, 105
Wakefield, Ruth, 75–76
Wald, George, 24
Walden Pond, 32, 66
Walker, David, 8, 29–30
Walker, Quock, 27
Walker's Appeal, 29–30
Waltham, Massachusetts, 79
Warren, John Collins, 54–55
Washington, Booker T., 51
Washington, George, 39
Waterman, "Bully," 98
Watson, James, 14
Wellesley College, 4
Wells, Horace, 54
Werner, Paul, 82
Wescott, W. Burton, 106
WFEB, *See* Worcester Foundation for Experimental Biology
Whirlwind computer, 125
Whitman, Massachusetts, 75
Whitney, Henry, 100
Wilcox, Russell, 108
Winthrop, John, 11, 18, 40–41
Woburn, Massachusetts, 69–70
Worcester, Massachusetts, 122
Worcester Foundation for Experimental Biology (WFEB), 81–82
Worcester Polytechnic Institute, 122
World Series, first, 104–5
World War I, 20, 74, 79, 87, 123
World War II, 1, 21–22, 56, 79, 86, 107, 123, 125
World Wide Web, 114–15
Wyeth, Andrew, 67
Wyeth, Jamie, 67
Wyeth, N.C., 67
Wyeth, Nathaniel J., 67

Y

Yale University, 5, 103
Young, Cy, 104

Z

Zuckerberg, Mark, 120–21

About the Authors

DR. ROBERT M. KRIM is the leading expert on the factors that have driven and continue to drive Massachusetts to be so innovative. As founder and leader of the Boston History and Innovation Collaborative, he put together a multi-university research team to comb through the region's four-hundred-year history and engage with hundreds of organizations and businesses to understand why nearly five hundred innovations that truly changed the nation or the world were developed in Greater Boston.

He has also developed a bus tour, enlivened by a talented actor, that explores the sites of many of Boston's most intriguing innovations, which inspired the creation of a children's book, *What's the Big Idea?*, as well as a play at the Boston Children's Museum. Most recently, his work on Boston innovation has been featured as a permanent exhibit at Boston's Logan Airport: *From Massachusetts to the World: Four Centuries of Innovation*.

Krim teaches innovation and entrepreneurship at Framingham State University, where he founded the Entrepreneur Innovation Center. For much of his professional life he started and ran two successful businesses. He has a BA from Harvard, studied economic history at London School of Economics, earned a masters in US History at University of California, Berkeley, a masters in economics, and a joint PhD/MBA in organization studies from Boston College. He is married and lives in Greater Boston.

ALAN R. EARLS is an author and independent writer covering business, science, and technology. During the 1980s he was an editor of *Mass High Tech* newspaper, reporting on the booming Greater Boston technology industry. He has also worked in the information technology and medical equipment industries in the region.

In addition to his consulting work with authors and businesses, he has written or cowritten several books, including *Route 128 and the Birth of the Age of High Tech*; *Raytheon Company: The First Sixty Years* (with Bob Edwards); *U.S. Army Natick Laboratories: The Science Behind the Soldier*; *Polaroid* (with Nasrin Rohani and Marie Cosindas); *Watertown Arsenal*; *Digital Equipment Corporation*; *Greater Boston's Blizzard of 1978*; and *Lucky Lamson in Her Own Words: The Story of the USS Lamson, DD-367*.

He was a guest curator for an exhibit on the rise of the Massachusetts technology sector at the Charles River Museum of Industry and presents regularly on historical and high-tech topics around New England. He lives southwest of Boston with his librarian-wife, Vicki, and their large cat.